Technology and Tools in Engineering Education

Technology and Tools in Engineering Education

Research and Innovations

Edited by

Prathamesh P. Churi

Vishal Kumar

Utku Kose

N. T. Rao

CRC Press is an imprint of the
Taylor & Francis Group, an **informa** business

First edition published 2022
by CRC Press
6000 Broken Sound Parkway NW, Suite 300, Boca Raton, FL 33487-2742

and by CRC Press
2 Park Square, Milton Park, Abingdon, Oxon, OX14 4RN

CRC Press is an imprint of Taylor & Francis Group, LLC

ISBN: 978-0-367-60774-6 (hbk)
ISBN: 978-0-367-60870-5 (pbk)
ISBN: 978-1-003-10229-8 (ebk)

DOI: 10.1201/9781003102298

Typeset in Times
by SPi Technologies India Pvt Ltd (Straive)

Dedication

This book is dedicated to all the Teachers, Students of Engineering and Researchers in Education Technology field

Contents

Preface

We are pleased to present you our first edition of book – Technology and Tools in Engineering Education: Research and Innovations. Education and Technology are two of the most important fields in any modern society because end-products and smart systems advance the quality of life and the welfare of people. Thus, providing quality education to the next generation will enable employability and contribution to the nation's development. This book will serve as a forum for the sharing existing and innovative practices in engineering education.

The book explores the innovative and research methods of teaching–learning process. The book focuses on the use of technology in the field of education. Technology can reduce the tremendous effort given by students to gather some printed book and journals for acquiring knowledge and increase students' focus on more important knowledge-gathering process. Equally important, technology can represent education in ways that help students understand latest concepts and ideas. The Education Technology also enables teachers to integrate project-based learning. With guidance from active teachers, students at different levels can use these tools to construct knowledge and develop skills required in modern society such as presentation skills and analytical skills. In the present time, the teacher's role in teaching is that of a facilitator. The teacher has to facilitate the learning by providing students with access to technology. The teachers can find the means to engage students more quickly in learning and to cater to the various needs of different students.

The book also focuses on providing opportunities for sharing and learning about innovative and best practices followed in the field of engineering education for enhancing the teaching–learning process. The book can provide a platform to academicians and educationalists to share out of the box ideas and best practices followed in teaching–learning process. In this book, participants can evolve new ideas, techniques, technologies in teaching methodologies, classroom tools and techniques, evaluation and assessment techniques and address the challenges to keep the millennial generation engaged in the teaching–learning process.

This edited book contains 10 chapters written by 22 contributors from 5 different countries. A short description of each chapter is given below.

Chapter 1 introduces the reader to the role of virtual laboratories in education. The teaching–learning aspects, pedagogy, concept demonstration, and the enriched learning experience made through virtual laboratories are discussed. The detailing of concepts, skills, programming aspects, and requirements of creating a virtual laboratory is elaborated. The types of virtual laboratories and gamification in virtual laboratories will also be explained in this chapter along with some illustrations of virtual laboratories for unconventional setup.

Chapter 2 discusses various issues, challenges, and solutions of classroom practice. Today, with the new advances in technology, there is a need to have innovative and active learning practices for Teaching. The proper learning at proper life span of a student plays a vital role in the mental as well as physical growth of that student. There are various types of learnings known as teaching pedagogies to have the two-way interactive sessions. The student's suggestions should be honored by giving them chance to express themselves and thus increasing the teaching–learning interactions. The teaching–learning sessions needs to be "Learner-centered" rather than "Teacher-centered" in classroom. Thus, it's the modern pedagogy that has learner diversity, where the Teacher has to cater to the diversity of learning styles and acts more like a Facilitator. The active learning pedagogies will definitely motivate and increase the active participation of the students in classroom.

Chapter 3 has Leveraging Information and Communication Technology (ICT) for Higher Education amidst COVID-19 Pandemic.

Chapter 4 Investigates academic transition during the COVID-19 pandemic: The case of an Indian Private University. In this chapter, author also presented the results of the impact of the transition to the online environment on academics at a private Indian University. The key findings indicate that the most popular tools adopted for communicating with students and colleagues in the work-from-home (WFH) situation were Zoom, WhatsApp, and Email. The study finds that during the pandemic, there was a clear increase in the adoption of Learning Management Systems (LMS) tools.

The objective of **Chapter 5** is to evaluate the effectiveness of computer-based assessment (CBA) and its use at different institutions of the world. Through CBA, students can be assessed in multiple dimensions, i.e. cognitive, meta-cognitive, and affective. In addition, CBA opens door for assessing students' complex form of knowledge, i.e. critical thinking, reasoning, etc. It keeps the track of students and teachers in terms of maintaining their portfolio, examination record, and path toward improvement in general. In this chapter, a systematic literature review has been carried out in which a number of articles have been reviewed in order to determine the effectiveness of CBA and its practices over the globe. It is found in literature review that CBA offers these advantages, including; time saving, auto-marking, provide quality and timely feedback, generating reports rapidly, reliability in results, economic, environmentally friendly, free from bias and cheating, and so on.

The purpose of **Chapter 6** is to analyze the transformation of ecological learning through ICT in the pandemic of COVID-19 in Pakistan. This study offers a great understanding about transformational shift of traditional classes toward ICTs. It will be helpful for the educators because it defines the gaps of using ICTs during Covid-19 situation.

Chapter 7 discusses the role of ICT in engineering education, Outcome-based Education (OBE) and its significance in the ever-changing educational context, dimensions, demands, challenges faced while OBE implementation in a traditional teaching–learning environment, use of ICT for analyzing and implementing complementary models such as ADDIE and addressing higher cognitive levels in all three contexts: design, delivery, and assessment.

Chapter 8 begins with an introduction to the crowdsourcing and scientific crowdsourcing and Theory of Planned Behavior (TPB). Then it presents methodology that covers search strategy and identification of sources. Finally, findings of systematic literature review are presented. Each section details a specific aspect of behavior of scientific workers, especially their attitude, subjective norms, and perceived behavioral control are presented.

Chapter 9 elaborates the mechanism of accreditation, tools involved in assessment, evaluation, and eventual attainment pertaining to the OBE concept. Also, the methodological effectiveness of various tools are discussed. A brief account of the accords in global and Indian philosophy covering the regulatory bodies are given. The merits and demerits of tools and technology use are also covered and discussed.

Chapter 10 describes innovative approach – 4QS Predictive Model based on machine learning for continuous Student Learning Assessment through rigorous results and analysis.

This book has an integrated approach of delivering the concept of engineering education through two aspects viz. technology and innovative tools. We hope that this book will create interest among the researchers who are working in similar fields.

– Prof. **Prathamesh P. Churi**, Assistant Professor, School of Technology Management and Engineering, NMIMS University, Mumbai, India

– Dr. **Vishal Kumar**, Assistant Professor, Assistant Professor, Bipin Tripathi Kumaon Institute of Technology, Uttarakhand, India

– **Utku Kose**, Associate Professor of Computer Engineering, Suleyman Demirel University, Isparta / Turkey

– **N. T. Rao**, Vice Chancellor, MIT World Peace University, Pune, Maharashtra, India

Acknowledgements

First and foremost, praises and thanks to God, the Almighty, for His showers of blessings throughout our work to complete the book – **T**echnology and **T**ools in **E**ngineering **E**ducation: **R**esearch and **I**nnovations (**TTEERI**) successfully.

Authors have been a real motivation and key in establishing CRC Press, Taylor, and Francis as one of the best books for publication in the subject of Data Protection and Privacy. We thank them all for considering and trusting our book for publishing their valuable work. We also thank all authors for their kind co-operation extended during the various stages of processing of the book in CRC Press, Taylor, and Francis.

For the success of any edited book, reviewers are an essential part and hence reviewers' merit sincere appreciation. The inputs of reviewers are used in improving the quality of submitted book chapters. The reviewing of a book chapter is essential to assure the quality of the chapter published in any book. We thank the following reviewers for their excellent contributions during the review process.

- Dr. Muhammad Mujtaba Asad, *Assistant Professor, Department of Education Management. Sukkur IBA University (SIBAU), Airport Road, Sukkur, Pakistan*
- Prof. Mukesh Soni, *Department of Computer Engineering, Smt S. R. Patel Engineering College, India-*
- Prof. Pankti Doshi, *Department of Computer Engineering, School of Technology Management and Engineering, NMIMS University, India*
- Prof. Radhika Chapaneri, *Department of Computer Engineering, School of Technology Management and Engineering, NMIMS University, India*
- Prof. Kamal Mistry, *Department of Computer Engineering, School of Technology Management and Engineering, NMIMS University, India*
- Dr. Shubham Joshi, *Department of Computer Engineering, School of Technology Management and Engineering, Shirpur, NMIMS University, India*
- Prof. Ameyaa Biwalkar, *Department of Computer Engineering, School of Technology Management and Engineering, NMIMS University, India*

The overwhelming responses from the authors across the world have been a real motivation and support in taking forward this book in the area of Data Protection and Privacy in the healthcare field. Last but not least, we would like to thank Ms. Cindy Renee Carelli (Executive Editor – Engineering, CRC Press, Taylor & Francis Group) and Ms. Erin Harris (Senior Editorial Assistant, Engineering, CRC Press, Taylor & Francis Group) for their valuable support in the book-editing process.

Editors

Prathamesh P. Churi (SMIEEE, MACM, LMCSI, LMISTE, MIGIP) is faculty member in Computer Engineering Department at School of Technology Management and Engineering, NMIMS University, India. He is senior member of IEEE. Currently, he is serving as associate editor of *International Journal of Advances in Intelligent Informatics* (Scopus Q4-Artificial Intelligence) and *International Journal of Innovative Teaching and Learning in Higher Education*. He is also the research mentor in Cerebranium. He is actively involved in peer-review process of reputed IEEE and Springer journals such as IEEE Transactions on Education, Springer Education and Information Technologies and 17 other journals. He has published 50+ research papers in National/International Conferences and Journals (Scopus, ESCI and SCI Indexed). He has four (Including one Australian patent) patents in the field of Wireless Sensor Networks, Machine Learning, and Internet of Things. He has edited three international books (CRC Press / Taylor & Francis Group publications) in the field of Data Privacy and Education Technology. He has been a keynote speaker, chair, convener in the international conferences including the flagship conferences like IEEE TALE 2017-2020, Springer ICACDS, etc. He recently received "Best Young Researcher award" by GISR Foundation for his research contribution in the field of Data Privacy and Security, Education Technology. He has also got appreciation award for best faculty from NMIMS University. He is active leader, coach, mentor, and volunteer in many non-profit organizations. He is also involved as a board of study member in many universities for curriculum development and educational transformations.

Recognizing his passion for research and teaching at a young age, he began teaching engineering students in core computer science branches at the age of 23. He started his career as a professor and researcher and has been working successfully in this field for the past 4 years. His relaxation and change involves pursuing his hobbies which mainly include expressing views in public by writing columns or blogs.

Vishal Kumar is working as an assistant professor in the Department of Computer Science & Engineering at Bipin Tripathi Kumaon Institute of Technology, Dwarahat, Almora, Uttarakhand, India. He also holds additional administrative responsibility as head, Central Computing Facility Centre & Head, Training & Placement Cell. He completed his bachelors from KEC Dwarahat, M. Tech from NIT, Hamirpur and PhD from Uttarakhand Technical University in Computer Science and Engineering, respectively. He has edited 5 books and published 23 research papers in IEEE Conferences/ SCI/ Scopus indexed journals of repute. He is an advisor/core team member/auditor in the International Federation of Green ICT. He has trained more than 450 Police Officers in the area of Cyber Security in his state (Uttarakhand). He holds a patent and completed (01) funded project. He is also serving as regional editor (Asia Pacific) of *International Journal of Forensic Software Engineering* published by Inderscience. He has organized more than 30 conferences and serves as the chairman of the steering committee for ICACCE/ICCCS/ICBDCI/ISCICT conferences worldwide under the flagship of India Research Group. He is also a member of 5G Working Group, Telecommunication Engineering Center, Govt. of India.

Utku Kose received the B.S. degree in 2008 in computer education from Gazi University, Turkey, as a faculty valedictorian. He received M.S. degree in 2010 from Afyon Kocatepe University, Turkey, in the field of computer science and D.S. / Ph.D. degree in 2017 from Selcuk University, Turkey, in the field of computer engineering. Between 2009 and 2011, he has worked as a research assistant in Afyon Kocatepe University. Following, he has also worked as a lecturer and vocational school vice director in Afyon Kocatepe University between 2011 and 2012, as a lecturer and research center director in Usak University between 2012 and 2017, and as an assistant professor in Suleyman Demirel University between 2017 and 2019. Currently, he is an associate professor in Suleyman Demirel University, Turkey. He has more than 100 publications including articles, authored and edited books, proceedings, and reports. He is also in editorial boards of many scientific journals and serves as one of the editors of the *Biomedical and Robotics Healthcare* book series by CRC Press. His research interest includes artificial intelligence, machine ethics, artificial intelligence safety, optimization, the chaos theory, distance education, e-learning, computer education, and computer science.

N. T. Rao is a vice chancellor of MIT WPU University, India. Dr. N. T. Rao has over three decades of result achieving experiences in the fields of education, training, strategic planning, accreditations, institutional governance, and other areas. Before joining MIT WPU, he was the Dean, Mukesh Patel School of Technology Management; Engineering (MPSTME), NMIMS University. During his career progression, he was associated with several institutions as an educator and professional engineer. Some of the organizations that he worked with include Intercontinental Consultants and Technocrats, New Delhi as a vice president; VIT University, Vellore as a Senior Professor, Director, International Relations; ABET Accreditation Officer; Government of Botswana as a principal roads engineer; NIT, Kurukshetra, as a lecturer and an assistant professor. Dr. Rao has introduced many initiatives during his stint at MPSTME resulting in the institution being ranked as the No. 2 Best Emerging Engineering Education Institution in India Today annual rankings in addition to debuting in NIRF rankings. He has strengthened and further streamlined academic and examination processes. He played a significant role in launching the *Engineering and Technology Review Journal* of the university. The crowning glory of his achievements at NMIMS was his spearheading the efforts for five engineering programs from scratch and readying them for the prestigious ABET Accreditation of USA. At VIT University, Vellore, he made significant contributions toward many facets of education helping the institution to catapult to the position of the No. 1 ranked private engineering college in India. As the director of International Relations, he expanded internationalization of engineering education and visited several prestigious institutions around the world to promote bilateral academic collaborations. He served on the Advisory Committee, the highest decision-making body of the University. He was University's NAAC Coordinator during the second reaccreditation process and helped university to achieve a score of 3.75 out of 4.0. Designing qualitative teaching–learning processes, he assumed the mantle of ABET Accreditation Officer and lead accreditation processes to make civil and mechanical engineering programs of VIT University as the first engineering programs in India to get the ABET accreditation.

Contributors

Amir A. Abdulmuhsin
University of Mosul
Mosul, Iraq

Kanwal Aftab
Sukkur IBA University, Airport Road
Sukkur, Sindh, Pakistan

Jayshree Aher
MIT World Peace University
Pune, India

Muhammad Mujtaba Asad
Sukkur IBA University, Airport Road
Sukkur, Sindh, Pakistan

Vishal Barot
LDRP Institute of Technology and Research
Gandhinagar, India

Zahid Hussain Bhat
AAA Memorial Degree College Bemina, Cluster University
Srinagar, India

Al-Karim Datoo
Sukkur IBA University, Airport Road
Sukkur, Sindh, Pakistan

Sharmishta Desai
MIT World Peace University
Pune, India

Prashant Gupta
Maharashtra Institute of Technology
Aurangabad, India

Zahid Husaain Khand
Sukkur IBA University, Airport Road
Sukkur, Sindh, Pakistan

Trishul Kulkarni
Maharashtra Institute of Technology
Aurangabad, India

Sachin Lal
Sir Padampat Singhania University
Udaipur, India

Ali Muhammad Mahar
Sukkur IBA University, Airport Road
Sukkur, Sindh, Pakistan

Regina Lenart-Gansiniec
Jagiellonian University
Kraków, Poland

Abhilasha Mishra
Maharashtra Institute of Technology
Aurangabad, India

Archana Nanade
Mukesh Patel School of Technology Management and Engineering, NMIMS University
Mumbai, Maharashtra, India

Sunny Nanade
Mukesh Patel School of Technology Management and Engineering, NMIMS University
Mumbai, Maharashtra, India

Himangi Pande
MIT World Peace University
Pune, India

B. Rajaguru
Maharashtra Institute of Technology
Aurangabad, India

Dilip Kumar Sharma
GLA University
Mathura, India

Fahad Sherwani
National University of Computer and Emerging Sciences
Karachi, Sindh, Pakistan

Upasana G Singh
University of KwaZulu-Natal
Durban, South Africa

Anita Thengade
MIT World Peace University
Pune, India

Bhagwan Toksha
Maharashtra Institute of Technology
Aurangabad, India

1 Virtual Experimentation
An Advanced Tool for Educational Technology

Prashant Gupta, Bhagwan Toksha, Trishul Kulkarni, B. Rajaguru, and Abhilasha Mishra
Maharashtra Institute of Technology, Aurangabad, India

CONTENTS

DOI: 10.1201/9781003102298-1

1.1 INTRODUCTION

Virtual laboratory is an alternative tool for physical experimental setup for conducting simulated experiments across various domains of Science, Technology, and Engineering (STE). The experiment is set up on a virtual platform, i.e. server and/or a website for users to access through the internet at any time and any place of the learner's convenience. It is a software for interactive learning based on simulations of real phenomena and consists of simulation programs, experimental units called objects that encompass data files, tools that operate on these objects and produce the intended results.

A virtual experimentation platform fabricated for distribution across the globe with the designed specification is no longer an impossible frontier to be conquered. In fact, it is the near future for design practice as well as education in virtual laboratories to be used as a substitute or supplement to the traditional laboratories (Bogusevschi et al., 2020; Ray & Srivastava, 2020; "Virtual Labs—The Future of Lab Classes", 2020). It is the role of instructors to reflect on how to make the best use of this emerging technology in teaching–learning process. This application of computer technology in the classroom environment has a significant role in enhancing the process. For instance, the use of artificial educational environments such as simulations in teaching and learning is increasingly becoming widespread. The virtual experimentation enhances students' skills in problem-solving, critical thinking, creativity, conceptual understanding, and their attitude in terms of motivation, interest, perception, and learning outcomes. These virtual laboratories allow the learner to "tinker" with laboratory equipment in the virtual space that behaves in almost the same way as it would in a real environment. The user performs a series of operations mimicking the actions in a physical laboratory procedure. These operations would yield authentic results and help in learning basic and advanced concepts through remote experimentation.

The virtual laboratories improve student performance when used as a learning tool. The virtual laboratory experiments are very effective when

(a) The reagents and equipment are expensive,
(b) Time requirement does not fit into the classroom schedule,
(c) There are ethical concerns,
(d) There is difficulty in result interpretation,
(e) There are issues in the handling of sophisticated instruments, and
(f) There is the involvement of hazardous material/s.

Virtual laboratories serve as an alternative solution for some of the problems related to classroom laboratory environments and prove effective in improving student performance in classroom education.

Today, most equipment has a computer interface for control and data storage. It is possible to design good experiments via the use of equipment that would enhance the learning of a student. Virtual laboratories are effective and realistic because it provides additional inputs to the students through various tools for learning, including web resources, video lectures, animated demonstrations, and self-evaluation. One of the major objectives of virtual laboratory is to serve as an alternative to costly laboratory equipment and resources, which are otherwise available to a limited number of users due to constraints on availability and geographical distances.

This chapter introduces, elaborates, and makes detailing of concept, philosophy, programming and requirements for virtual laboratory aiming to provide the relevant information in creation or implementation of distance and e-learning in STE disciplines. The role of virtual laboratory in education, teaching–learning aspects, pedagogy and role of the content creator, types of virtual laboratories, gamification aspects, misconceptions, concept cracking, developer role, programming aspects, and unconventional virtual laboratories will be discussed.

There is a challenging question in front of us before we go ahead and know more about virtual laboratories. "What are the aspects of a good instructional environment and are these aspects being addressed by virtual laboratories?" We shall try and see if it gets addressed going forward.

1.2 ROLE OF VIRTUAL LABORATORY IN EDUCATION

It has been proven that the traditional teaching methods which are teacher-centric do not give effective results in learning process which is due to the fact that curriculum is not student centered (Shaw et al., 2019). The enhancement of students' individual capabilities, intelligence, and creative thinking can only be achieved through student-centered learning methods (Odabaşı, 1997). A virtual laboratory is an alternative to traditional learning where students can interact and practice their course content in a much livelier and enriching manner. Online education means not only learning the course curriculum but also enhancing the cognitive ability of the learner, which in turn enables the better implementation of theories into practice.

The recent literature suggests a number of emerging technologies appearing in concern with the future of developing education especially for the teaching and learning of STE domain. Some of the developed concepts are novel while others are mere adaptations of existing ideas (Villar-Zafra et al., 2012). The most relevant technological examples include e-learning, distance learning, virtual laboratories, virtual worlds and virtual reality, dynamics-based virtual systems, Internet of Things, and the new emerging conceptualized technologies of immersive education that integrates many of these ideas together. The reason for this variation lies in the fact that these fields often require laboratory exercises to provide effective skill acquisition and hands-on experience (Potkonjak et al., 2016).

For the online teaching of experimentation, many researchers have come together via this challenging concept of immersive teaching and learning (Queiroz et al., 2019) and the Immersive Learning Research Network (iLRN) was evolved (*Immersive Learning Research Network – iLRN*, n.d.). In the developing education system, concepts such as distance learning are widely used for enhancing teaching and learning.

The new emerging technologies such as virtual space augmented reality, computer graphics, and computational dynamics can overcome some of the potential difficulties in the up-growing technology of e-learning.

One of the strengths of Virtual learning is that laboratory exercises are an integral part of courses in STE (Potkonjak et al., 2010). Therefore, it is mandatory for every student to acquire skills as a part of the learning process. There are two ways in which distance learning can accommodate these aspects, i.e. remotely triggered laboratories and simulation-based laboratories (Torres et al., 2016). Virtual experimentation encompasses the introduction of new concepts as a part of learning, which supports more constructive education and training activities in complex engineering scenarios.

However, the individualization in performing a virtual laboratory experiment along with remote access may encourage a reduction in team activities, discussion and hinder learning by each other's experience between teacher–learner and learner–learner interactions. This may lead to an incomplete assessment of the affective domain. These virtual setups may also result in hindering student learning with real equipment and physical instrumentation setup. This may lead to an incomplete assessment of the psychomotor domain. The alluring nature of virtual laboratories should not make one completely replace the real-world labs as the students may miss out on learning by doing with their own hands, an experiential learning feature important to be honed during STE education. Also, the designing of a virtual laboratory is a challenge and may require tremendous thinking to really translate it to the level of physical lab setup. If not done, it may lead to plagiarism during the assessment as all may end up getting the same readings/results (Chan & Fok, 2009).

1.3 TEACHING ASPECTS

Teaching is a complicated process and is not as straightforward as it appears at the first sight as there are many facets to it that combine to make learning happen. The parameters like content knowledge, quality of instruction, teaching climate, classroom management, responsive feedback, student engagement, efficient use of lesson time, coordinating classroom resources and space are perhaps the necessary parameters for good teaching–learning. Besides having a deep knowledge of the subject that an instructor possesses, effective communication of content to their students, understanding the way students think about the content, evaluating the thinking behind students own learning methods, and identifying common misconceptions of the students about the content for the learning to be better is necessary. The quality of instruction includes teachers being skilled in effective questioning and the use of assessment. The instructors may also deploy techniques that consist of reviewing previous learning, giving adequate practice time to the students, and judging student grasping power. The quality of the teaching–learning relationships between teachers and students is also very important. When done well, teachers scaffold student learning by progressively introducing new skills and knowledge.

A good instructional environment creates a climate that constantly demands more and pushes students to succeed. A good instructional environment challenges students, develops a sense of competence, attributes success to effort rather than ability, and most importantly values resilience to failure. To achieve meaningful learning and/or high-level knowledge, revisiting the concept plays a vital role. Out of various definitions, learning can be understood as a change in long-term memory. An important concern is to make sure that the students truly learn, rather than merely experience, the curriculum which can be addressed by revisiting knowledge after a gap (space in time). It works because, after a gap, students have to work harder to retrieve knowledge from their memory. This effort means that they are more likely to remember the information in the longer run. A way to achieve this is to introduce planned and dedicated spacing. One of the ways to achieve this is to make the learning tools available more often to the users which don't restrict the learning to only four walls of the classroom. One such tool, i.e. virtual laboratory, can help this activity of retrieving the content besides the quizzes, tests, presentations by the students; and it can be made more interesting with its integration in the process of learning.

One of the hidden challenges the instructor community faces is adapting to the teaching skills the way they were taught to them at least a decade back. The simplification of research that has been carried out in this context exhibits that revisiting the sources of teaching knowledge and prior experience plays an important role in shaping faculty teaching practices. One of the most prominent approaches being followed is "Faculty teach the way they were taught" (Cox, 2014; Üstünbaş, 2017). There are far too many changes in the technology related to any domain to stick to the same pattern of teaching. This statement above need not to be necessarily true but can help us in exploring the teaching methodologies and the way ahead. The elaborations and/or explanations that a teacher employs in teaching may include knowledge derived from the experiences as an instructor, student, researcher, industrial, and other non-academic roles. One of the key factors that enrich teaching skills are the different ways of presenting the same content to improve learning by bringing out the creativity of the teacher. Figure 1.1 describes the conceptual framework of development and

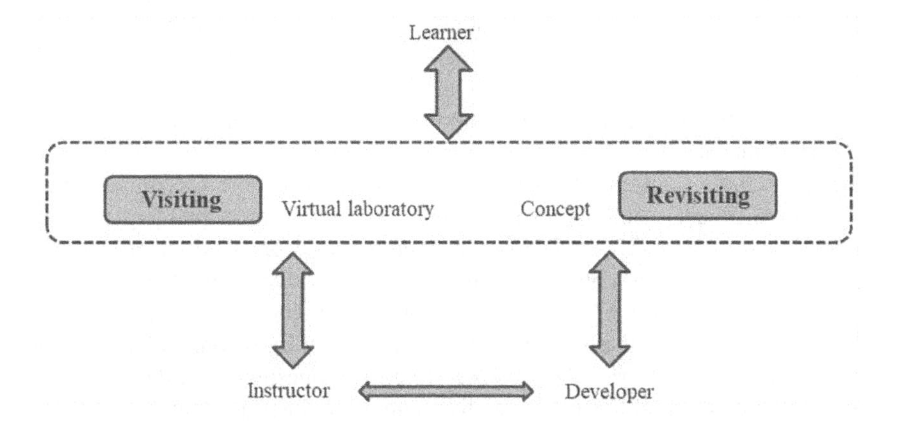

FIGURE 1.1 Visiting and revisiting concept in different roles as learner, instructor, developer.

usage of virtual experimentation. The three roles with which the virtual experimentation interacts are the following:

1. Developers who are involved in the development of virtual experimentation;
2. Instructors intended to use the available experiments as a part of their instructions; and
3. Users or students using virtual experimentation for their learning.

Visiting and revisiting concepts in different roles as a learner, instructor, and developer are interrelated processes and contribute toward creating a better learning environment.

The parameters of concern in physical (real-world) laboratory experiments are that these may be expensive, time-consuming, and occasionally constrained by safety concerns. The laboratory courses as a supplement to classroom lectures are often the first classes to be reduced from the scope of curriculum to minimum required only for adhering to the guidelines or prescriptions of apex advisory bodies. This is a sorry situation because several theoretical science courses benefit from an experimental counterpart. The challenge for us as teachers is to create engaging, relevant, and personalized learning experiences for all learners. The latest equipment and consumables are often so expensive, making it difficult for institutions to provide students with access to modern equipment. With these constraints, the students may develop a general perception of learning content as non-motivating and feel disconnected from the real world. The students in many to most of the cases find the course content taught in such a conventional classroom setup being "boring" or "very boring".

1.4 LEARNING ASPECTS

Let us explore the aspects of learning and try to relate them in a virtual experimentation environment. The three basic aspects of learning are learning by listening, watching, and doing. These aspects should contribute to developing the skill set of learners in cognitive, affective, and psychomotor domains such as skills in inquiry, problem-solving, reasoning, creative thinking, information processing, evaluation, self-awareness, motivation, communication, social along with managing and empathy. The three aspects of learning could very well be addressed by virtual experimentation as it keeps learners involved and engaged along with providing an individual experience for them to navigate and complete the set tasks addressing the "learning by doing" dimension.

Virtual experimentation allows students to explore the topic by comparing and contrasting different scenarios to pause and restart application enabling reflection and taking notes at their own learning pace along with practical experimentation experience. It is a learner's playground for experimentation which allows learners to apply what they are about to learn in a real-world setting long before they ever could in the real world. Also, the data analysis, comparison, discussion, and conclusions become feasible. The learners are exposed to critical questions and they are left to their own devices to find the answer with an interactive interface that is supposed to walk them through the problem scenario until a solution is determined.

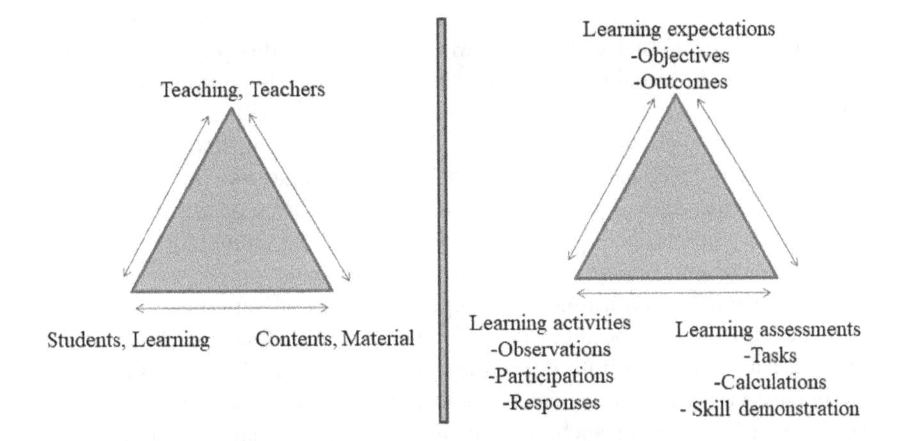

FIGURE 1.2 Learning triangles as reported in the literature and designed in the context of virtual experimentation for meaningful learning.

For meaningful learning to occur, the focus of the teacher remains on how to connect three points of the triangle, i.e. student learning, content, and teaching to design a good learning environment with appropriate instructions. These three were established as three corners of the triangle in the conceptual framework by Murata et al. and the team investigated them for use as representations in teaching multi-digit subtraction and behavioral effects occurred over time. This idea when extended to virtual experimentation could further help in understanding student learning in terms of observations, responses, their participation, and the learning expectations set in terms of objectives, outcomes, and assessments set in a virtual environment. Figure 1.2 depicts the learning triangle as adopted from the literature (Murata et al., 2012) and designed in the context of virtual experimentation for meaningful learning.

1.5 PEDAGOGY

In general, the definition of pedagogy is understood as the approach to teaching, thinking, and practices of educators with virtues of accompanying learners in the learning process, caring about them, and bringing learning to life. It also refers to the theory and practice of learning and influential interactions between the process and the social, political, and psychological development of learners. Shulman (Shulman, 1987) quoted that "Teaching necessarily begins with a teacher's understanding of what is to be learned and how it is to be taught". The mechanism which teachers employ for learning to occur is called pedagogy. In the report titled "Innovating Pedagogy 2019, Exploring new forms of teaching, learning, and assessment, to guide educators and policymakers", it was marked that the educator community is growing confidence with reference to the usage of virtual studios for teaching–learning, and they recognize the educational value of virtual studios and new avenues offered by them in pedagogy (Ferguson et al., 2019).

The integration of technology into pedagogy is the basic concept of virtual experimentation and thus can be understood and explored with the help of a research article published about Technological Pedagogical Content Knowledge providing a teacher framework in educational technology (Mishra, 2019). In this pioneering work, the interplay among three main components of learning environments, i.e. content, pedagogy, and technology, was described. Virtual experimentation is a classic example of incorporating authentic design-based activities for teaching technology as shown in Figure 1.3. The understanding of connections between platforms to hold the experiment, coding, interactions (technology), the subject matter (content), and the methods of teaching it (the pedagogy) would let the teaching community and students dive in details of philosophy regarding virtual experimentation.

The experiments with a time scale dependency can be varied and cut down on time scale with virtual experimentation. For example, microbial growth and decay in a physical laboratory setup need 24 hours to complete. The Geiger Muller counter experiment in nuclear physics needs readings to be taken at an interval of 30–120 seconds with hundreds of such readings to be taken. This situation is handled by resetting the time scale and accelerating the experiment by reducing the time lapse steps. The other side to this is the phenomenon that is so fast on a natural time scale that it is difficult to visualize the process and such processes could be slowed down

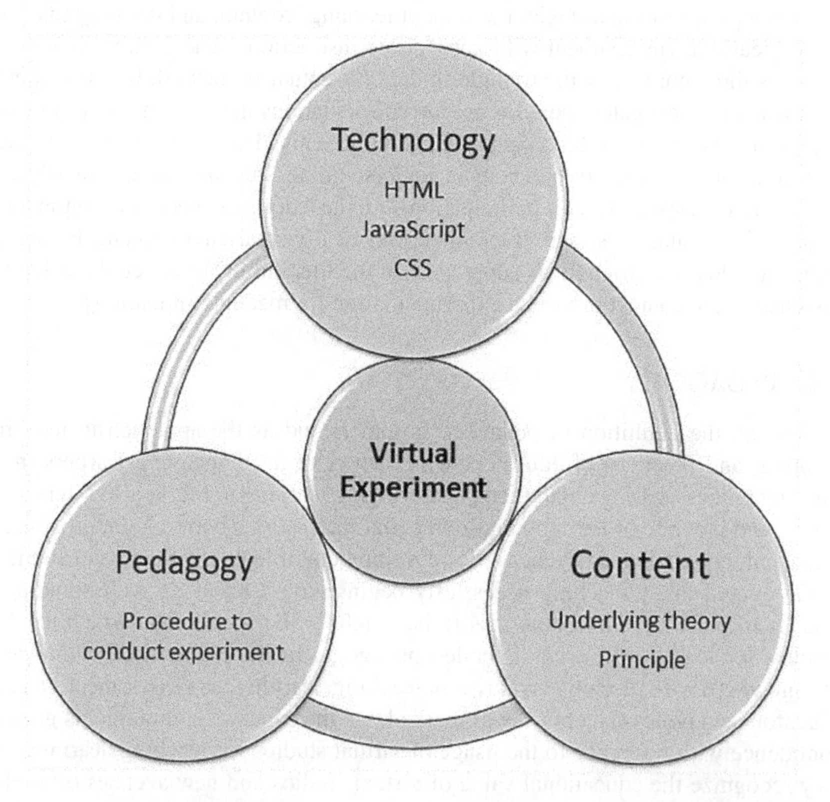

FIGURE 1.3 Virtual experimentation as a concenter of three components, namely technology, pedagogy, and content designed for instructors for integration of technological, pedagogical and content knowledge.

to enable the user to go in further detailing and increase the detailed understanding of the process.

The pedagogical potential of the teaching process of programming and impact of virtual laboratories on learning with incorporation of virtual programming in learning management system was reported by Marílio Cardoso (Cardoso et al., 2017). A case study on virtual laboratory usage in chemistry-related courses found that there are positive effects of virtual experimentation on students' cognitive and affective learning as compared to traditional teaching methods (Tüysüz, 2010). A similar study was reported with using a virtual laboratory in chemistry along with the merits/demerits of using virtual laboratories and the user perception about effectiveness and helpfulness in learning. The users were reported to be performing better in cognitive domains with respect to the experimental process and results with precise use of the engagement time (Tatli & Ayas, 2010). In an article reporting about the effectiveness of virtual laboratory structure in Computer Engineering, authors brought in important conclusions about the teacher–student interaction. It was reported that to ensure the success of the virtual laboratory implementation in the teaching–learning process, human and pedagogical factors are equally important as students do value these factors along with the technological factors considering the communication environment as an essential resource (Prieto-Blazquez et al., 2009). João Batista Bottentuit Junior et al. studied the role of mobile devices in learning virtual laboratory experimentation and suggested that the use of mobile devices could lead to an improvement in the user experiences. They also inferred that the adaptation of mobile technology in virtual experimentation will expedite the usage of e-contents to a virtual mobile laboratory (Junior et al., 2007).

The teaching–learning in the case of twin heat exchanger experiments was reportedly enhanced via the use of a virtual environment as a tool. The use of flexible teaching and training reportedly achieved overcoming the challenges imposed by the hardware, minimizing the effort needed by laboratory support staff and chances of errors/failures along with avoidance of mechanical repetitions (Murphy et al., 2002). The effectiveness of a virtual laboratory for helping students transfer engineering theory to the design and building of a model truss was studied (Karweit, 2002). It was highlighted that graphics-based design tools provided as a supplement in teaching–learning were able to greatly enhance the learning by filling the knowledge gap between abstract mathematical concepts and physical understanding. Cecilia Chan and Wilton Fok carried out a systematic case study in Electrical and Electronic Engineering on learning experiences in virtual laboratory training with a motivation to analyze whether virtual experimentation could emerge as a substitute for traditional laboratory training by providing an equivalent and comparable learning experience for students. In this study, it was concluded that the virtual laboratory as an instructional tool and the delivery of its learning outcomes is promising, thereby, demanding more research to fully exploit virtual experimentation as a delivery medium and its related tools in the enhancement of teaching and student learning (Chan & Fok, 2009).

With the features as discussed above, the virtual experimentation evidently addresses the users' cognitive functions in a better way by attracting attention, engaging the learner, sustaining motivation, exposing students to detailed and advanced skill sets, and improving them by addressing the psychomotor domain of learning.

Estriegana and the team employed the technology acceptance model (TAM) for virtual laboratories and other online learning environments (OLE) to study students' acceptance of technology and the process of adopting online learning. They developed a theoretical model with the help of the TAM framework to understand the various individual factors associated with users' acceptance of OLE learning resources. They considered various factors such as attitude, behavioral intention to use, efficiency, perceived ease of use, perceived usefulness, playfulness, and Satisfaction while using OLE resources. The study was conducted for the first-year students enrolled for the degree program of Computer Engineering and Information Systems taking the course Fundamentals of Computer Technology. The sample size consisted of 223 students (189 males and 34 females). It was concluded that efficiency and playfulness are the most important factors that influence students' adoption of virtual laboratories and other OLE's. The term efficiency, in particular, refers to the ability to accomplish a task with the minimum expenditure of time and effort whereas playfulness is an inherent belief or motive which shapes the individual's experience of the environment. To sum it up, it was recommended that virtual tools such as virtual laboratories should be designed with a playful approach to improve motivation and acceptance of these tools (Estriegana et al., 2019).

1.6 MISCONCEPTIONS IN LEARNING

"Effective Teaching" is more than just the transmission of information from teacher to students and has the scope of holistic development of the learners at all levels of education. The students create their own understanding of the topic in their mind aside from what and how something is being taught which may lead to differences in understanding at the cognitive level. One important factor contributing to this difference is that the student occasionally creates simplistic mental models when trying to make sense of complex issues and stick to these conceptions which are rather "misconceptions", also referred to as "non-scientific ideas" and "alternative conception". Students' misconception about the concepts is a vital concern as it interferes with the learning and therefore is needed to overcome. Some of the reported examples of such misconceptions include natural selection (Abraham et al., 2009), the concept of mass (Stamenkovski & Zajkov, 2014), and ozone depletion problem (Groves & Pugh, 2000). It is essential to identify misconceptions about the concept or topic carefully. The help to achieve this could be fetched from the literature available or could be easily figured out by conducting open-ended discussion about the topic among the students.

The students may hold on to these misconceptions about the concepts even after classroom instructions or performing physical laboratory experiments, if there is no direct mechanism which specifically points out and clarifies it. Virtual experiments can play a vital role in such situations, wherein a specific instance can be planned for students to commit the "mistake" and learn through it, which ultimately help in the elimination of the misconception. There are scientific reports available which have reported the use of virtual experiments, being useful in the removal of misconceptions (Abraham et al., 2009; Schneps et al., 2014; Wibowo et al., 2016; Zhou et al., 2011). Abraham et al. reported the use of interactive simulated laboratory for effectively dispelling some common misconceptions about natural selection. Schneps et al. reported that even brief exposures to virtual 3D simulations driven by a pinch-to-zoom display of the solar system were very helpful in addressing student misconceptions about the scale of space and time in astronomy. Zhou et al. reported the use of an interactive virtual environment in addressing the misconceptions among the students related to force and motion. Wibowo et al. insisted that their findings strongly support that the microscopic virtual mediacan be used as an alternative instructional tool in order to help students by confronting them, resulting in the reduction of their scientific misconceptions and developing an understanding of physics concepts.

1.7 ROLE OF CONTENT DEVELOPER

The content developers are considered to be the backbone for creating a virtual laboratory as they play a pivotal role in the journey from concept visualization to portraying it on a computer screen. For the virtual laboratory experiment to attain desired outcomes, a great deal of knowledge and thought process is involved in designing the experiment. The strategy of the developer has to be goal-oriented so that the users can inherit knowledge through experiential learning. There should be a clear passage of the information that the users are expected to ace through the learning process. There are several components for the learning of a virtual laboratory experiment such as text for theoretical understanding, animation, and exposition of steps involved in carrying out the experiment, converting results into plots for analysis, questionnaire based on the goal, i.e. learning outcomes, and performing the experiment in a two- or three-dimensional environment which is virtual in nature (Yu et al., 2005). The factors required for the development of good content regarding virtual laboratory experiments are shown in Figure 1.4.

1.7.1 PEDAGOGY AND BLOOM'S TAXONOMY

The teaching–learning process is cognition-based and it becomes imminent that any process which involves teaching–learning has to be based on the same. Such a teaching–learning process focuses on effective and efficient use of brain functions dealing with the same to gain knowledge and understanding through experiential, sensual, and thoughtful learning. An American educational psychologist Benjamin Bloom and his coworkers in 1956 published the classification of educational goals

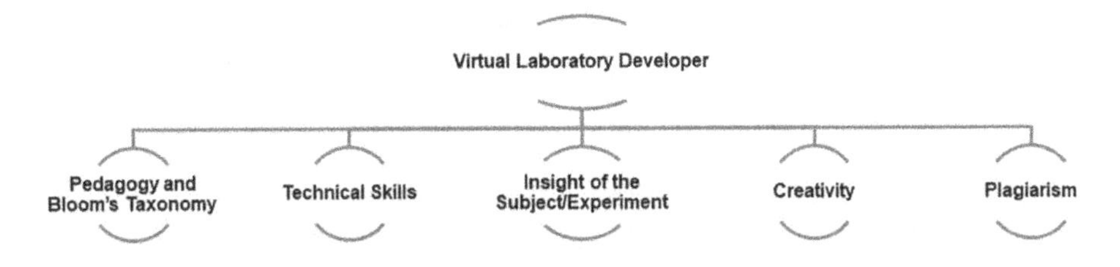

FIGURE 1.4 Requirements for Content Development.

as a taxonomy which is named after him and is referred as "Bloom's Taxonomy" (Bloom, 1956). However, it was revised and published by a group of instructional researchers, curriculum theorists, and testing and assessment specialists in 2001 (Anderson et al., 2001). The focus of the new content was on the dynamic classification of the older concept. The fact that some tasks are more difficult than others and are tougher to complete makes sequential learning all the more important for designing the tasks in teaching–learning process. The tasks have to be designed in steps as a part of the virtual laboratory experiments (easy to tough) with respect to the following hierarchy, i.e.

1. Recalling facts and basic concepts (Remember);
2. Explaining concepts (Understand);
3. Using information in new situations (Apply);
4. Drawing connections among ideas (Analyze);
5. Justifying a decision (Evaluate); and
6. Producing novel/original work (Create).

1.7.2 Technical Skills

The technical skills required for content development is largely governing two aspects, i.e. designing and coding. Several free and open-source graphic editors are available for both 2D images/designs and 3D modeling. A summary of the same is given in Table 1.1. The coding-end is largely dedicated to making and using various

TABLE 1.1
Overview of various free and open-source software used for design activities related to creating content for virtual laboratory experiments (*Design Labs-OnlineLabs.in-Virtual Laboratory Simulations for Science Education,* n.d.)

Serial No	Software	2D/3D	Use
1	Google SketchUp	3D	Engineering Design
2	Blender	3D	Content Creation such as modeling, sculpting, rendering, realistic materials, rigging, animation, compositing, video editing, game creation, and simulation
3	DAZ 3D	3D	Modeling, accessories, and environment
4	Google Web Designer	3D	Animations and engaging creatives
5	GIMP	2D	Photo manipulation, basic graphic creation, and illustration
6	Inkscape	2D	Vector graphics editor with similarities in work to Adobe Illustrator or CorelDraw
7	My Paint	2D	Drawing tablet-centric expressive drawing and illustration tool
8	Project Dogwaffle	2D	Drawing, sketching, and animating
9	Irfanview	2D	Graphic viewer and editor
10	Meesoft	2D	Image analyzer

frameworks such as JavaScript or its superset typescript which is also used in making interactive web pages. The advantage with typescript is the consistent assigning of language primitives such as string number and object. It also has the advantage of providing highlight in the code at places of unexpected behavior which also lowers the chances of bugs. Furthermore, the use of free and open-source software such as Atom and Brackets can be employed as source code editors for coding and development.

1.7.3 Insight of the Subject/Experiment

It is always important from a developer perspective to have knowledge in the subject/experiment that is being developed. For instance, a cross-functional team will always be employed to make a flight simulator and will consist of an expert in trade, i.e. a certified pilot with similar experience. If the experiment for which the virtual laboratory is being designed has been conducted by the developer and/or conducted under the keen supervision of the developer (possibly a faculty/tutor in the same stream), the content developed for designing the virtual laboratory experiment would be very good and technically accurate. Also, in such a scenario if the lab is simulator based, the developer can include accurate representative results.

If the content being developed is from a course/subject which falls under the expertise/interest area of the developer, there can also be an inclusion of results derived from research work published on the experiment being performed. For example, in a Fourier Transform Infrared Spectrophotometry (FTIR) experiment, to study the material through its spectral image which is obtained after scanning the material to be tested, a library of FTIR images consisting of several sample images can be incorporated in the experiment from published research work and can be coded to club with the samples being tested virtually. The developer can also include the aspects of the experimentation which often remains hidden/unexplored during the actual physical laboratory experiment. A virtual experiment for learning nuclear fission or testing a sample for infectious diseases caused due to coronavirus (Covid 19) gives the flexibility to the user for exploring areas which will remain unexplored in physical setups. The user can learn by committing mistakes which he/she will not be allowed to do in an actual environment.

1.7.4 Creativity

Creativity is integral in development of a virtual laboratory experiment as the physical setup would not be available as a part of instructions/interactions, and the user will be using a computer interface with peripherals such as keyboard and mouse to perform the experiment which is not the usual practice in the majority of the laboratories such as chemistry, biology, physics, etc. The visualization of physical experimental setup and putting it in the form of 2D images and/or 3D models is important for the user to have a feel of the experimental environment. Furthermore, as the eventual aim is to enhance the teaching–learning process, outcomes from conducting the experiment, i.e. goals have to be decided and tasks/activities have to be designed and incorporated in a pedagogical sequence which makes the learning activities start from lower levels and move toward higher levels of cognition.

1.7.5 Plagiarism

It is a moral obligation of the academicians and educators to comply with ethics in intellectual content as in today's world of multiple search engines almost all the information is at a single click of the mouse. As a virtual laboratory is an intellectual technology product, plagiarism is an important aspect that the developer has to keep in mind while doing its creation. The plagiarism in a virtual laboratory can creep into the content in the form of images used to depict experimental setups, animations, or simulations to demonstrate the working of the experiment, text used to illustrate basic theoretical concepts required for the experiment, etc. It is advisable to draw images and make animations/simulations with the software list given in Table 1.1.

1.8 TYPES OF VIRTUAL LABORATORY

The virtual laboratories can be classified into simulation-based and remotely triggered laboratory.

1.8.1 Simulation-Based Laboratory

These laboratories consist of experiments modeled with the use of varying levels of mathematical equations. The importance of the internet in such a case is prominent as the simulations are remotely done at a high-end server and conveyed to the user accessing the experiment via its use. The simulation-based setups have been used since the use of flight simulation program "Link Trainer" back in 1928 to train thousands of pilots who fought World War II (Feisel & Rosa, 2005). These type of laboratories are significant as they carry a potential to imitate actual live situations during experimentation, are portable, easy to use/access, and are cost-effective (Hong Shen et al., 1999). Also, its employability over a large scale is another feature that makes it useful for simultaneous use within a large community of users. A typical simulation-based laboratory in the area of biotechnology has been exhibited by Radhamani et al. (Radhamani et al., 2015). These laboratory experiments are enabled with simulations to allow the students to receive training from a distance from the actual experimental setup and gaining exposure in terms of practical knowledge. Moreover, the most important aspect of learning through these laboratories is the "learning by doing mistakes" concept which one cannot afford with the real instrument/equipment setup. This enhances user learning as some parameters often remain unexplored while always doing the right thing in order to keep the real setup working and intact. However, in spite of its equivalence to physical lab for explaining and reinforcing theoretical concepts, a simulation-based laboratory does not exactly reflect the real-world situation as it exhibits limited comparability for experimentation (Striegel, 2001).

1.8.2 Remotely Triggered Laboratory

The experiments to be conducted in these laboratories are triggered from a remote location and experimental output is relayed back to the user over the internet. The access to remote experiment is heavily dependent upon client–server architecture.

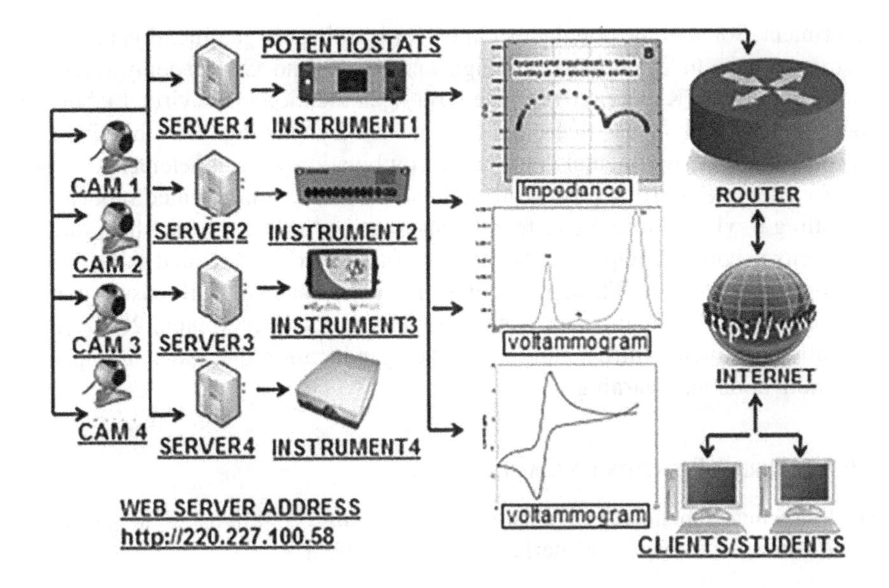

FIGURE 1.5 Setup of Remotely triggered electrochemistry laboratory for distance learners (Saxena & Satsangee, 2014).

The significance to the user is of conducting real-time experiments from anywhere irrespective of its geo-location with respect to the physical setup. These laboratories can serve real experience distance learning purposes with minimal setup costs. The cost of actual experimental equipment is not incurred as only computer and high-speed internet facility is required. s These are omnipresent in today's world due to their versatile nature and usefulness in day-to-day activities. The working mechanism of a typical remotely triggered lab has been exhibited in the field of robotics as shown in Figure 1.5 (Nayak et al., 2014). However, there are limitations in terms of its simultaneous usage as it can only accommodate limited users, and availability to multiple users needs proper scheduling and management.

There are three important essentials (Saxena & Satsangee, 2014) for the server-side of a remotely triggered laboratory, i.e.

- Sensor which is responsible for remote monitoring of the experiment;
- I/O hardware responsible for mediation between instrumental setup and software by adjusting input signals and measuring/generating output; and
- Software for remotely connecting and conducting measurements in the equipment.

1.9 GAMIFICATION IN VIRTUAL LABORATORY

For the purpose of improvement of user engagement and motivation, gamification has been employed by teachers for day-to-day classroom activities. Also, it is a very popular and important aspect while designing a good virtual laboratory experiment. Some of the popular examples for learning via gamification in a virtual environment include Duolingo (foreign languages) and Codecademy (programming languages) (Knautz, 2015). The term "gamification" in a virtual laboratory doesn't necessarily mean learning experiments by playing. It rather indicates toward learning experimental concepts in combination with the elements of gaming, for example, starting a two-roll mill and preparing a masticated rubber sheet by cutting it with a knife for re-feeding and mixing, an activity in which the user will perform with the help of accessing psychomotor domain-related skills through the computer interface. These game elements or mechanisms will ensure engagement and user participation for achieving these tasks (Zichermann, 2011). There are various elements (direct and indirect) of gamification that have an impact on user motivation and learning.

1.9.1 Direct Elements of Gamification

The following elements are responsible for visual (direct) stimulus for action and are hence known as Direct User Interface (UI) Elements.

1.9.1.1 Narrative Elements

A positive backstory and a plot are crucial to the success of learning from virtual laboratory experiment which enables placement of the application framework (Kapp, 2012; Sailer et al., 2013). Furthermore, onboarding (induction/trial) process can be employed for introducing the user with the mechanics of the application so that the user may be able to achieve the targets defined by the backstory with assigning new tasks (Zichermann & Cunningham, 2011).

1.9.1.2 User Profile

A user profile is necessary for a simulated setup so that the user can relate to the virtual experimental environment. For example, if a chemistry experiment is to be performed virtually, a user profile may incorporate features of a human body (facial features, height, body physique etc.) to select and assemble a humanoid character performing the experiment with an apron, safety glasses, goggles, gloves, shoes, and other personal protective equipment as per the laboratory requirements. It is more of a personalized approach toward learning so that the user can relate to the backstory and there is a considerable increase in his/her interest and motivation (Frery et al., 2002).

1.9.1.3 Tasks

These are the activities and objectives that are meant to be completed to achieve the set goals. It can also be in the form of quests and missions inside an experimental plot. The achievements can be incentivized upon completion which makes the laboratory simulator more compelling and intensifies user engagement due to the human urge to achieve more. If the user fails, provision of multiple attempts (wherever possible) may be made available so that the user tries a different approach on the subsequent attempt which thereby invigorates the learning process (Kapp, 2012).

1.9.1.4 Stars/Ratings

Stars/Ratings can be termed as feedback on how the user is performing. It is a reflection of how good or bad the learning process is happening via the virtual experiment. There can be an increase in the number of star ratings based on the level of difficulty as it is with some competitive exams such as GRE, GMAT, etc. Even time can be recorded and can be simulated to be a parameter for giving ratings depending upon how quickly the task was done.

1.9.1.5 Levels and Badges

Levels and badges are interconnected and persuade users to do the tasks better which are possible by an efficient combination of the cognitive and psychomotor learning processes. As discussed before, the various levels can be spread over Bloom's Taxonomy where "Remember" is the easiest level whereas "Create" is the toughest. Badges are also a type of rating scale just like giving stars which is visible to the users at the time they start. It may also contribute to increase in user awareness of their study habits and make them work on improving the same (Hakulinen et al., 2015).

1.9.1.6 Ranking Lists

Ranking lists make the users competitive in nature which might bring out the best in them for enhancing the learning process. The user may go through the list to see the rank allotted individually or in group which might also do team building along with learning (Sailer et al., 2013).

1.9.2 Indirect Elements of Gamification

These are elements which are linked with the direct ones to enhance their performance. Some of them include connecting user objectives with laboratory objectives (Salen et al., 2004) which in turn helps the user in connecting with the virtual experiment, feedback mechanisms (Deci & Ryan, 2009; Deterding et al., 2011) which should be constructive and influential for the motivation of the user in a sense that a binary code of right or wrong can be assisted with engaging hints, advice, and tips so that the user may improve the performance and creating competition in collaborative structures which challenges the users on fronts such as coordination, creation, and cooperation (Lewis et al., 2016).

1.10 PROGRAMMING IN VIRTUAL LABORATORY

A simulator is a software or set of instructions written in programming languages and shared with users by various means like websites, learning management systems, or by sharing through physical memory storage devices. The simulators are having a set of control to convert real-world actions into a virtual environment. The models mimicking the real-world situations in terms of simulations with possible interactions are created in the form of simulators. Simulations are used for explorations of mathematical models, creating graphics, images, videos, and so on. The tools that can be employed for creating a virtual laboratory simulator shall be discussed. Some of the paid/free and open-source tools available can be broadly classified for client-side coding and server-side coding, for example, Scilab-xcos, LTspice, Simulink, LabVIEW, Flash coding, etc. to name a few.

Scilab-xcos, LTspice, Simulink, and LabVIEW are software that need installing in computer system for working and creating simulations. The features offered by the software listed above are summarized in Table 1.2.

Furthermore, the approaches in the development of virtual experiment from a viewpoint of resource sharing can be Client-side coding (HTML, JS, and CSS) and

TABLE 1.2
A list of features offered by software for coding in Virtual Laboratory

Sr. No.	Name of Software	FOS/ Commercial	Features	References
1	Scilab-Xcos	FOS	• High-level programming language and graphics editor that enables physical modeling to design hybrid dynamical system models. • Effectively used by engineers and scientists for numerical computations, mathematical operations and analysis, graphical analysis for signal processing, basic circuit simulations, dynamic systems, and interfacing with hardware.	(Jain, 2016)
2	LT spice	FOS	• An analog electronic circuit simulation software for analog device design and simulation wherein the results can be represented in graphical and text formats. • Widely used for schematic capturing and viewing waveforms with enhancements and models for easing the simulation of analog circuits.	(Guru Prasad, 2016)
3	Simulink	Commercial	• A MATLAB-based graphical programming environment that can be employed for mathematical modeling and analyzing various parameters by using mathematical solver options. • Used for model-based design and multi-domain simulations. • Capable to generate a code for real-time implementation and interfacing with hardware like DSP processors, microcontrollers, and raspberry pi.	(Dwan & Bechert, 1993)
4	LabVIEW		• Design software for industrial automation, instrument control, and data acquisition system. • Used in simulators with test-, measurement- and control-related requirements as they offer rapid access to hardware and data insights.	(Higa et al., 2002; Yi et al., 2005)

Server-side coding (PHP and JS) that are tools used for making better virtual laboratories and have the capability to run on all available web browsers.

1.10.1 Flash Coding

Flash is graphics-based animation file to create and save. The flash code was developed by Macromedia and its coding can be done in HTML5, CSS, and JavaScript to develop a simulator. Flash code in HTML5 doesn't require any additional plugin and can be made to run on various devices and browsers with the only requirement being having flash in the system from which the simulator is run (Huang et al., 2010). However, the support of flash will be withdrawn by Adobe, the parent company with End of Life in December 2020 as announced in July 2017 (*Adobe Flash Player End of Life*, 2017).

A simple image coded in HTML 5 is shown in Figure 1.6. The image is a pictorial representation of a scooter at the stationary position which will move from the left side to the right side in the image to reach its final position. The HTML code can be written on notepad++ or any HTML source code editor. The requirements for the development of such a simulator include basic HTML knowledge, background image, and scooter image/gif file.

1.10.2 Client-Side Coding

This program runs on a client machine or browser and involves working with user display/interface and/or any miscellaneous processing like reading/writing cookies that might happen on the client system. A developer should have the basic knowledge of any of the programming languages such as HTML, AJAX, VB Script, JavaScript,

FIGURE 1.6 A pictorial representation of the starting stage of animation for movement of scooter moving from left side to right side.

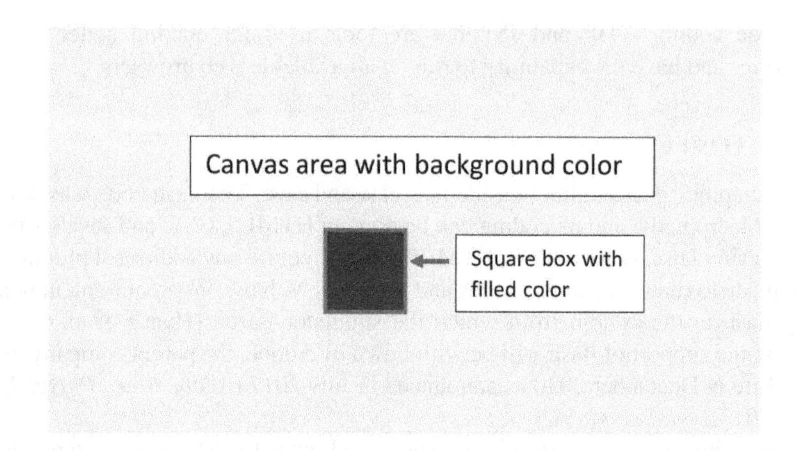

FIGURE 1.7 Canvas area and square box created by HTML coding for virtual laboratory screen.

and CSS (Vishesh et al., 2017). A simple example to quote involves the creation of a specific area for a simulator which is known as canvas wherein images, videos, and animations can be added. Also, its size, i.e. height and width, can be configured based on the experimental simulator setup with the addition of canvas border and background color, and so on. The inputs can be received and processed in this area to make interactions happen. However, the records of interactions in terms of data generated occur at local memory storage temporarily. The canvas area and square box filled with color are shown in Figure 1.7.

1.10.3 Server-Side Coding

The programming languages that can be used for server-side programming are PHP, C++, JSP, Java, Python, etc. As shown in Figure 1.8, server-side coding has the virtues of a program which takes care of the web page content generation and interacts with a server in terms of querying. It performs operations over the database, access,

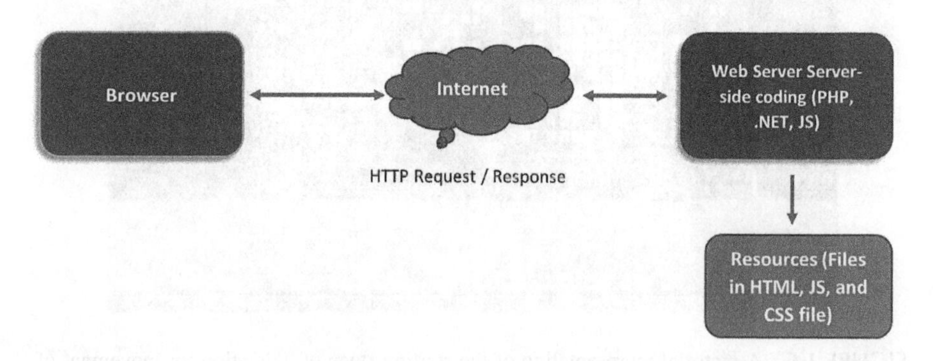

FIGURE 1.8 A simple block diagram of server-side coding.

and interact with own/other servers. Such interactions in a virtual laboratory occur by the users accessing the virtual laboratory experiment through the use of a browser on the internet (Ortiz, 2014).

1.11 DEVELOPING VIRTUAL LABORATORY IN A THEORETICAL FRAMEWORK

This section elaborates on how the understanding of Blooms levels and the TPACK model can help the developer or instructor to develop a virtual experiment and incorporate it in the teaching–learning activities. A virtual experiment, the simulator display of which is shown in Figure 1.9, was selected and developed for the course "Engineering Physics" offered at the first year of engineering. In this experiment, when a monochromatic source of light is allowed to fall on an optical element, i.e. diffraction grating, the light beam undergoes bending and forms a series of bright spots on the screen on both sides of the central spot. The student is expected to determine the wavelength of the light source by measuring the distance between diffraction grating and screen and taking down the first-order, second-order spacing, and so on.

In terms of technical specifications, the behavior of various elements was controlled by JavaScript; Hypertext markup language (HTML) is used for basic structuring and Cascading Style Sheets (CSS) are used for presenting the experimental elements on the screen. The learning expected from the experiment was cut down or dissected in terms of various learning objectives during the planning phase of the development. These learning objectives were then transferred into tasks to be performed by the student. The hierarchy of tasks was set in an appropriate ordering as per Bloom's Taxonomy. There were tasks for identifying the experimental objects addressing the "Remember" level followed by tasks and interactions set at the

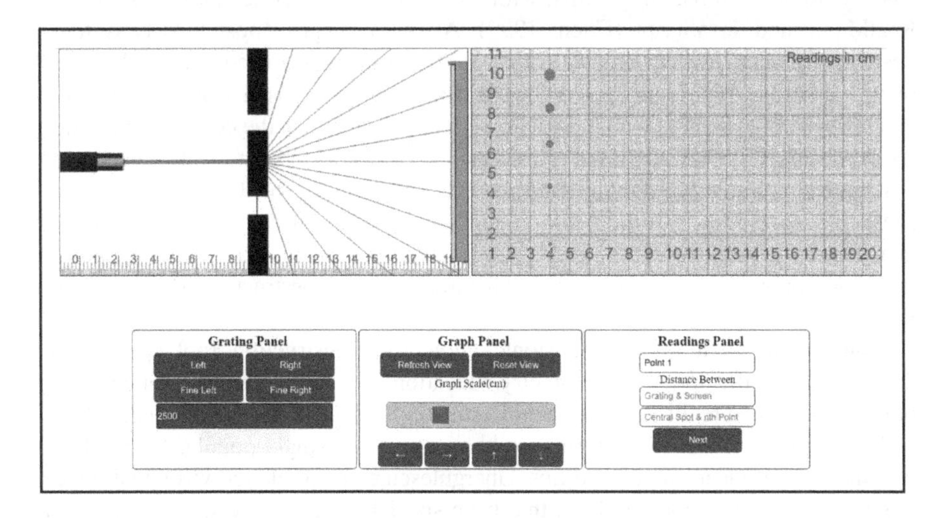

FIGURE 1.9 A view of the simulator window for a virtual laboratory developed on the experiment entitled "Wavelength of laser by using diffraction grating".

"Understand" level. The task to apply experimental conditions, record the observations in a table, and calculation was set at the "Apply" level. The results found out by the student and the standard results were made available to the student for comparison and analysis in terms of accuracy and percentage error at the "Analyze" level. The concept checkpoints were carefully designed and were incorporated for making the learning meaningful. In general, no more than a couple of diffraction gratings would be available in a physical laboratory for students to do the experiment. However, in this virtual experiment, there are 5 sets of gratings with a variation provided in the number of rulings. One of the major concepts to understand in this virtual experiment is the fact that X-ray diffraction differs from the diffraction of visible light. The student will learn it through an activity via this experiment. Also, an experimental understanding of the wavelength equation regarding no impact on observations due to the source moved closer or far is reflected via an activity that can be done as a part of the designed virtual experiment.

1.12 VIRTUAL LABORATORY FOR UNCONVENTIONAL LABORATORY SETUP

Generally, laboratory experiments are associated with the STE courses, but with virtual environment, the idea of "virtual laboratory" can be extended to cover the broad areas like mathematics, language learning, and communication skills. Computer-assisted language learning (CALL) (Levy, 1997) is one of the approaches to language teaching–learning which make use of computers and information technology. Software or applications used in a CALL environment are designed specifically for learning of foreign/second language or adapted for such purposes. The faculty at National Taiwan Ocean University have developed a virtual language lab for helping students to develop listening skills, with the help of new web technologies such as RealMedia and JavaScript (Chen, 1999). At initial stages, most CALL systems designed for language learning were kind of tutorials that consisted of a combination of introduction, traditional lessons, word drills, scenario-based lessons/dialogues, and series of questions to which the learner responds and then the computer provides the feedback. With the advances in artificial intelligence and natural language processing technologies, CALL systems are becoming intelligent. University of Aizu, Japan has developed "WordBricks project"—an intelligent computer-assisted language learning virtual environment which is intended to serve as a "'virtual language lab" that supports open experiments with natural language constructions (Mozgovoy & Efimov, 2013).

Construct3D is a three-dimensional geometric construction tool specifically designed for mathematics and geometry education. It is based on the mobile collaborative augmented reality system "Studierstube" (Kaufmann & Schmalstieg, 2002).

Furthermore, virtual laboratories are becoming increasingly popular for teaching the students with learning disabilities. Ghergulescu et al. employed virtual laboratories for secondary school students that have special education needs (Ghergulescu et al., 2018). The results analysis of pictorial Torrance tests of creative thinking exhibited that the students' creative thinking has improved significantly in terms of various mental characteristics such as fluidity, flexibility, and originality. Adamo-Villani et al. developed a new immersive 3D learning environment to increase mathematical skills of deaf children (Adamo-Villani et al., 2006).

1.13 CONCLUSION

A detailed overview of existing concepts and technologies in the field of software-based virtual laboratories was discussed to identify current trends and to contribute a solid foundation for future research and development in the application and use of virtual laboratories. In the future, the vision is to overcome the barriers that still prevent the wide-scale implementation of e- and distance-learning, through the use of virtual worlds, immersive education, and other technologies pertinent to STE disciplines. The virtual laboratories are going to play a vital role in education as they touch upon teaching–learning aspects which are sometimes difficult to achieve in a physical classroom/laboratory setup. With the development of computational facilities in education, there is a gradual shift of teaching–learning activities online as they offer a host of advantages over conventional learning methods. Also, the issue with the availability of tools for practical learning due to various reasons in setting up of necessary infrastructure, unforeseen situations where teaching–learning activities are limited to online mode and ease in usage of technology via mobile phones/laptops/desktops are leading to usage and Research &Development of virtual laboratories. While designing teaching–learning activities through virtual laboratories, a greater emphasis has to be given to pedagogy, misconceptions, and concept cracking. The gamification of virtual laboratories while doing content creation necessitates incorporating elements that are responsible to increase user engagement and motivation. We have also discussed coding aspects and virtual laboratory for unconventional setup as a part of this chapter. Although the role of the developer is pivotal in the creation of a virtual laboratory, feedback from the academic community and users/students is important as in any other human–computer interface for enhancement of teaching–learning model proposed by virtual laboratories.

Though experts believe that virtual lab systems are the desired initial step in STE education, there is no replacing the role of physical laboratories in training at an advanced level. The need of hands-on experience with real equipment is inevitable, but with the increasing development in virtual technology, the gap between the real and virtual world is shrinking in terms of transfer of knowledge and skills.

REFERENCES

Abraham, J. K., Meir, E., Perry, J., Herron, J. C., Maruca, S., & Stal, D. (2009). Addressing undergraduate student misconceptions about natural selection with an interactive simulated laboratory. *Evolution: Education and Outreach*, *2*(3), 393–404. doi:10.1007/s12052-009-0142-3

Adamo-Villani, N., Carpenter, E., & Arns, L. (2006). An immersive virtual environment for learning sign language mathematics | ACM SIGGRAPH 2006 Educators program. *Siggraph '06*, 20-es. doi:10.1145/1179295.1179316

Adobe Flash Player End of Life. (2017, July). https://www.adobe.com/in/products/flashplayer/end-of-life.html

Anderson, L. W., Krathwohl, D. R., & Bloom, B. S. (2001). *A taxonomy for learning, teaching, and assessing: A revision of Bloom's taxonomy of educational objectives* (Complete ed.). Longman.

Bloom, B. S. (1956). Taxonomy of educational objectives. Vol. 1: Cognitive domain. *New York: McKay*, *20*, 24.

Bogusevschi, D., Muntean, C., & Muntean, G.-M. (2020). Teaching and learning physics using 3D virtual learning environment: A case study of combined virtual reality and virtual laboratory in secondary school. *Journal of Computers in Mathematics and Science Teaching*, *39*(1), 5–18.

Cardoso, M., Barroso, R., Vieira de Castro, A., & Rocha, Á. (2017). Virtual programming labs in the computer programming learning process, preparing a case study. *EDULEARN17 Proceedings*. 7146–7155. doi:10.21125/edulearn.2017.2704

Chan, C., & Fok, W. (2009). Evaluating learning experiences in virtual laboratory training through student perceptions: A case study in Electrical and Electronic Engineering at the University of Hong Kong. *Engineering Education*, *4*(2), 70–75. doi:10.11120/ened.2009.04020070

Chen, H.-J. H. (1999). Creating a virtual language lab: An EFL experience at National Taiwan Ocean University. *ReCALL*, *11*(2), 20–30. doi:10.1017/S0958344000004924

Cox, S. E. (2014). *Perceptions and influences behind teaching practices: Do teachers teach as they were taught?* [Brigham Young University—Provo]. https://scholarsarchive.byu.edu/etd/5301/

Deci, E., & Ryan, R. (2009). The "What" and "Why" of goal pursuits: Human needs and the self-determination of behavior. *Psychological Inquiry*, *11*(4), 227–268.

Design Labs-OnlineLabs.in-Virtual laboratory simulations for science education. (n.d.). Retrieved August 21, 2020, from http://onlinelabs.in/design

Deterding, S., Dixon, D., Khaled, R., & Nacke, L. (2011). From game design elements to gamefulness: Defining "gamification". *Proceedings of the 15th International Academic MindTrek Conference: Envisioning Future Media Environments*, 9–15. doi:10.1145/2181037.2181040

Dwan, T. E., & Bechert, T. E. (1993). Introducing SIMULINK into a systems engineering curriculum. *Proceedings of IEEE Frontiers in Education Conference—FIE '93*, 627–631. doi:10.1109/FIE.1993.405447

Estriegana, R., Medina-Merodio, J.-A., & Barchino, R. (2019). Student acceptance of virtual laboratory and practical work: An extension of the technology acceptance model. *Computers & Education*, *135*, 1–14. doi:10.1016/j.compedu.2019.02.010

Feisel, L. D., & Rosa, A. J. (2005). The role of the laboratory in undergraduate engineering education. *Journal of Engineering Education*, *94*(1), 121–130. doi:10.1002/j.2168-9830.2005.tb00833.x

Ferguson, R., Coughlan, T., & Egelandsdal, K. (2019). *Innovating pedagogy 2019: Open university innovation report 7. Milton Keynes: The Open University.* (No. 9781473028333; pp. 1–41). https://iet.open.ac.uk/file/innovating-pedagogy-2019.pdf

Frery, A., Kelner, J., Moreira, J., & Teichrieb, V. (2002). User satisfaction through empathy and orientation in three-dimensional worlds. *CyberPsychology & Behavior* doi:10.1089/109493102761022878

Ghergulescu, I., Lynch, T., Bratu, M., Moldovan, A.-N., Muntean, C. H., & Muntean, G.-M. (2018). STEM education with atomic structure virtual lab for learners with special education needs. *EDULEARN18 Proceedings*, *1*, 8747–8752. doi:10.21125/edulearn.2018.2033

Groves, D. F., & Pugh, D. A. (2000). Cognitive Illusions as Hindrances to Learning Complex Environmental Issues. *Reports-Research*, 9.

Guru Prasad. (2016). *Circuit simulation examples using LTspice*. https://www.researchgate.net/publication/303565325_Circuit_Simulation_Examples_using_LTspice

Hakulinen, L., Auvinen, T., & Korhonen, A. (2015). The effect of achievement badges on students' behavior: An empirical study in a university-level computer science course. *International Journal of Emerging Technologies in Learning (IJET)*, *10*(1), 18–29.

Higa, M. L., Tawy, D. M., & Lord, S. M. (2002). An introduction to LabVIEW exercise for an electronics class. *32nd Annual Frontiers in Education*, *1*, T1D-T1D. doi:10.1109/FIE.2002.1157905

Huang, Q., Lin, S., & Abdel-Ghaffar, K. (2010). Flash Coding Scheme Based on Error-Correcting Codes. *2010 IEEE Global Telecommunications Conference GLOBECOM 2010*, 1–6. doi:10.1109/GLOCOM.2010.5683364

Immersive Learning Research Network – iLRN. (n.d.). Retrieved August 26, 2020, from https://immersivelrn.org/

Jain, S. (2016). Evaluation of scilab on basic operations for research and teaching. *International Journal of Research in Computer Applications and Robotics.*, *4*(2), 7–11.

Junior, B., Batista, J., & Clara, C. (2007). Virtual laboratories and M-learning learning with mobile devices. *Proceedings of International Milti-Conference on Society, Cybernetics and Informatics*, 275–278.

Kapp, K. M. (2012). *The gamification of Learning and instruction: Game-based methods and strategies for training and education*. John Wiley & Sons.

Karweit, M. (2002). *Enhanced learning through a "virtual laboratory."* 7.

Kaufmann, H., & Schmalstieg, D. (2002). Mathematics and geometry education with collaborative augmented reality. *ACM SIGGRAPH 2002 Conference Abstracts and Applications*, 37–41. doi:10.1145/1242073.1242086

Knautz, K. (2015). *Gamification in University Didactics—Conception, Implementation and Evaluation of a Game-Based Learning Environment* [Dissertation, Institute for Language and Information» Information Science]. https://docserv.uni-duesseldorf.de/servlets/DocumentServlet?id=36429

Levy, M. (1997). *Computer-assisted language learning: Context and conceptualization*. Oxford University Press.

Lewis, Z., Swartz, M., & Lyons, E. (2016). What's the point? A review of reward systems implemented in gamification interventions. *Games for Health Journal*, *5*(2), 93–99. doi:10.1089/g4h.2015.0078

Mishra, P. (2019). *Considering contextual knowledge: The TPACK diagram gets an upgrade* (Vol. 35). Taylor & Francis.

Mozgovoy, M., & Efimov, R. (2013). WordBricks: A virtual language lab inspired by Scratch environment and dependency grammars. *Human-Centric Computing and Information Sciences*, *3*(1), 5. doi:10.1186/2192-1962-3-5

Murata, A., Bofferding, L., Pothen, B. E., Taylor, M. W., & Wischnia, S. (2012). Making connections among student learning, content, and teaching: teacher talk paths in elementary mathematics lesson study. *Journal for Research in Mathematics Education*, *43*(5), 616–650. doi:10.5951/jresematheduc.43.5.0616

Murphy, T., Gomes, V. G., & Romagnoli, J. A. (2002). Facilitating process control teaching and learning in a virtual laboratory environment. *Computer Applications in Engineering Education*, *10*(2), 79–87. doi:10.1002/cae.10011

Nayak, S., Vakrani, P., Purohit, A., & Prasanna, G. N. S. (2014). Remote Triggered Lab for Robotics: Architecture, Design and Implementation Challenges. *2014 IEEE Sixth International Conference on Technology for Education*, 214–217. doi:10.1109/T4E.2014.24

Odabaşı, F. (1997). Eğitimde Sistem Yaklaşımı ve Eğitim Teknolojisi. *Eğitim ve Bilim*, *21*(106).

Ortiz, A. (2014). Server-side web development with JavaScript and Node.js (abstract only). *SIGCSE '14*. doi:10.1145/2538862.2539001

Potkonjak, V., Gardner, M., Callaghan, V., Mattila, P., Guetl, C., Petrović, V. M., & Jovanović, K. (2016). Virtual laboratories for education in science, technology, and engineering: A review. *Computers & Education*, *95*, 309–327. doi:10.1016/j.compedu.2016.02.002

Potkonjak, V., Vukobratović, M., Jovanović, K., & Medenica, M. (2010). Virtual Mechatronic/Robotic laboratory – A step further in distance learning. *Computers & Education, 55*(2), 465–475. doi:10.1016/j.compedu.2010.02.010

Prieto-Blazquez, J., Herrera-Joancomarti, J., & Guerrero-Roldán, A.-E. (2009). A virtual laboratory structure for developing programming labs. *International Journal of Emerging Technologies in Learning (IJET), 4*(0), 47–52. doi:10.3991/ijet.v4s1.789

Queiroz, A. C. M., Nascimento, A. M., Tori, R., & da Silva Leme, M. I. (2019). Immersive Virtual Environments and Learning Assessments. In D. Beck, A. Peña-Rios, T. Ogle, D. Economou, M. Mentzelopoulos, L. Morgado, C. Eckhardt, J. Pirker, R. Koitz-Hristov, J. Richter, C. Gütl, & M. Gardner (Eds.), *Immersive Learning Research Network* (pp. 172–181). Springer International Publishing. https://doi.org/10.1007/978-3-030-23089-0_13

Radhamani, R., Sasidharakurup, H., Kumar, D., Nizar, N., Achuthan, K., Nair, B., & Diwakar, S. (2015). Role of Biotechnology simulation and remotely triggered virtual labs in complementing university education. *2015 International Conference on Interactive Mobile Communication Technologies and Learning (IMCL)*, 28–32. doi:10.1109/IMCTL.2015.7359548

Ray, S., & Srivastava, S. (2020). Virtualization of science education: A lesson from the COVID-19 pandemic. *Journal of Proteins and Proteomics, 11*(2), 77–80. doi:10.1007/s42485-020-00038-7

Sailer, M., Hense, J., Mandl, H., & Klevers, M. (2013). Psychological perspectives on motivation through gamification. *Interaction Design and Architecture(s) Journal, 19*, 28–37.

Salen, K., Tekinbaş, K. S., & Zimmerman, E. (2004). *Rules of play: Game design fundamentals*. MIT Press.

Saxena, S., & Satsangee, S. P. (2014). Offering remotely triggered, real-time experiments in electrochemistry for distance learners. *Journal of Chemical Education, 91*(3), 368–373. doi:10.1021/ed300349t

Schneps, M. H., Ruel, J., Sonnert, G., Dussault, M., Griffin, M., & Sadler, P. M. (2014). Conceptualizing astronomical scale: Virtual simulations on handheld tablet computers reverse misconceptions. *Computers & Education, 70*, 269–280. doi:10.1016/j.compedu.2013.09.001

Shaw, T. J., Yang, S., Nash, T. R., Pigg, R. M., & Grim, J. M. (2019). Knowing is half the battle: Assessments of both student perception and performance are necessary to successfully evaluate curricular transformation. *PloS One, 14*(1), e0210030.

Shen, H., Xu, Z., Dalager, B., Kristiansen, V., Strom, O., Shur, M.S., Fjeldly, T.A., Lu, J.Q. & (1999). Conducting laboratory experiments over the Internet. *IEEE Transactions on Education, 42*(3), 180–185. doi:10.1109/13.779896

Shulman, L. (1987). Knowledge and teaching: Foundations of the new reform. *Havard Educational Review, 57*, 1–21.

Stamenkovski, S., & Zajkov, O. (2014). Seventh grade students' qualitative understanding of the concept of mass influenced by real experiments and virtual experiments. *European Journal of Physics Education, 5*(2), 20. https://doi.org/10.20308/ejpe.13383

Striegel, A. (2001). *Distance education and its impact on computer engineering laboratories. 31st Annual Frontiers in Education Conference. Impact on Engineering and Science Education. Conference Proceedings (Cat. No.01CH37193), 2*, F2D-4. doi:10.1109/FIE.2001.963707

Tatli, Z., & Ayas, A. (2010). Virtual laboratory applications in chemistry education. *Procedia—Social and Behavioral Sciences, 9*, 938–942. doi:10.1016/j.sbspro.2010.12.263

Torres, F., Candelas, F., Puente, S., Pomares, J., Gil, P., & Ortiz, F. (2016). Experiences with virtual environment and remote laboratory for teaching and learning robotics at the University of Alicante | Publons. *International Journal of Engineering Education, 22*(4), 766–776.

Tüysüz, C. (2010). The effect of the virtual laboratory on students' achievement and attitude in chemistry. *International Online Journal of Educational Sciences, 17*, 37–53.

Üstünbaş, Ü. (2017). "I TEACH GRAMMAR AS I WAS TAUGHT": IS IT THE VIEW OF ELT TEACHERS IN TURKEY? *International Journal of Language Academy*, *5*(18), 8–16. doi:10.18033/ijla.3601

Villar-Zafra, A., Zarza-Sánchez, S., Lázaro-Villa, J. A., & Fernández-Cantí, R. M. (2012). *Multiplatform virtual laboratory for engineering education. 2012 9th International Conference on Remote Engineering and Virtual Instrumentation (REV)*, 1–6. doi:10.1109/REV.2012.6293127

Virtual Labs—The Future of Lab Classes. (2020, August 13). *Silicon India*. https://hr.siliconindia.com/career-news/virtual-labs-the-future-of-lab-classes-nid-213621.html

Vishesh, S., Nandan, A. S., Kavya, P. H., Izhar Ahmed, M., Gayathri, K., Keerthana, B., & Ganashree, H. R. (2017). Server-side and client-side scripting of "Shutter Speed", *IJARCCE*, 6(9). https://ijarcce.com/upload/2017/september-17/IJARCCE%2042.pdf

Wibowo, F. C., Suhandi, A., Rusdiana, D., Ruhiyat, Y., & Darman, D. R. (2016). *Microscopic Virtual Media (MVM) in Physics Learning to Build a Scientific Conception and Reduce Misconceptions: A Case Study on Students' Understanding of the Thermal Expansion of Solids. Proceedings of the 2015 International Conference on Innovation in Engineering and Vocational Education.* 2015 International Conference on Innovation in Engineering and Vocational Education, Bandung, Indonesia. doi:10.2991/icieve-15.2016.52

Yi, Z., Jian-Jun, J., & Shao-Chun, F. (2005). A LabVIEW-based, interactive virtual laboratory for electronic engineering education. *International Journal of Engineering Education*, *21*(1), 94–102.

Yu, J., Brown, D. J., & Billett, E. E. (2005). Development of a virtual laboratory experiment for biology. *The European Journal of Open, Distance and E-Learning*, *8*.

Zhou, S., Han, J., Pelz, N., Wang, X., Peng, L., Xiao, H., & Bao, L. (2011). Inquiry style interactive virtual experiments: A case on circular motion. *European Journal of Physics*, *32*(6), 1597–1606. doi:10.1088/0143-0807/32/6/013

Zichermann, G. (2011, April 26). *The purpose of gamification—O'Reilly Radar*. http://radar.oreilly.com/2011/04/gamification-purpose-marketing.html

Zichermann, G., & Cunningham, C. (2011). *Gamification by design: Implementing game mechanics in web and mobile apps* (1st ed.). O'Reilly Media, Inc.

2 Classroom Practices
Issues, Challenges, and Solutions

Jayshree Aher, Sharmishta Desai, Himangi Pande, and Anita Thengade
MIT World Peace University, India

CONTENTS

2.1 INTRODUCTION

Learning is a practice of acquiring value-based education. Today it's not only the academics that matters, but physical, cognitive and socio-emotional aspects too need to be imbibed in the students. The new advances in technology have innovative and active learning practices for Teaching. Now, it's the student-centric Education era. As quoted by John Holt, "*Learning is not the product of teaching but Learning is the product of the activity of Learners*". The proper learning at proper life span of a student plays a vital role in the mental as well as physical growth of that student. There are various types of learnings known as teaching pedagogies to have the two-way interactive sessions. The world is changing and we have to adapt to these altering situations for having effective learning skills with appropriate outcomes. There must be a thoughtful process for the classroom teaching. The student's suggestions should be honored by giving them chance to express themselves and thus increasing the teaching–learning interactions. In such type of teaching pedagogy the teachers too should learn from the students. The students may have doubts, which can be cleared within the class by discussing with Faculty and other students. With the need

DOI: 10.1201/9781003102298-2

of time for a faculty the role is changing from Teacher to Facilitator by utilizing various strategies and available tools to facilitate effective teaching–learning interactions between these elements. The improved thinking makes the understanding of concepts easier, thus resulting in good inferences and conclusions. As the students are techno savvy, their habit of using smart phones can be properly channelized for more active learning.

2.2 CHALLENGES AND ISSUES

Teaching practice is one of the integral part for a teacher's teaching and learning activity. John Foncha et al. with their research study examine some of the difficulties faced by the teachers and students during professional teaching practice, such as technical–nontechnical resources, placement activities, learners discipline, classroom management, etc. It also discusses the reflectivity approach of the students–teachers during the exposure to the learning environment to gather the theoretical aspects. According to an article by teachers (March 2019), there are various challenges and issues to be handled during actual teaching process. Revision of the courses/ syllabus with the new demands of society and industry is a need to cope with the current technology and trend. Accordingly access to the updated content with innovative educational trends requires proper planning which may be time consuming. Along with teaching, building good statistical reports too become a challenge with excessive paper work. With the outcome-based education, proper learning and feedback is an important task. As the number of students per session increases, the learning styles of each learner/student may differ. The teacher should use different possibilities for explaining a concept. This can include live demonstrations, real-time case studies, simulations, etc. Family background too plays an important role in getting the education and choosing the career path. At times it may happen that the student is the first one in family to have such an educational experience. Thus, having counseling sessions and mentoring the learner is also necessary. Use of teaching pedagogies will help to overcome the challenges of classroom practices. The social development and advances in technology have led to the need to have effective learning practices. It has thus demanded for the multiple insights coming from different interdisciplinary educational values. Having too much of a burden of learning may distract the learner. The new teaching–learning pedagogies and formative assessment strategies take into account the prior knowledge and experience of the students. It indeed helps to simplify the understanding and learning of the individual. This develops the intrapersonal skills and interpersonal communication for the learner.

2.3 ACTIVE LEARNING

The useful teaching–learning sessions needs to be "*Learner-centered*" rather than "*Teacher-centered*" in classroom. The *Student* is our ultimate and the most important stakeholder. The active learning in classroom must include the student involvement with the Faculty acting as a Facilitator rather than Teacher. As shown in Figure 2.1, Facilitator should facilitate the teaching–learning process in the classroom to have dialogue sharing.

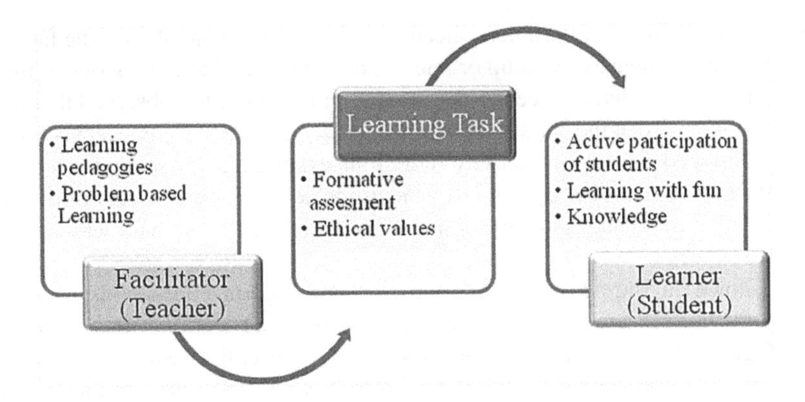

FIGURE 2.1 Learner-centered teaching process.

This will also improve the communication skills of the students minimizing the stage fear in some of the minds. The ethical and moral values in the education will help the students to have and shape the correct path in their lives. The advantage in answering questions immediately after the discussion of a topic in the session is that it makes the learning simpler and easier. It helps in reinforcing and revising the concepts. Curiosity leads to questions and solving these queries has the outcome as new research and innovation. In physical classroom the Facilitator may facilitate a topic and have revision sessions. One of the good things of structured group discussions, forums is to have focused discussions for a particular concept or module. It increases the involvement of the student's interaction by improving their decision power.

The National Education Policy (NEP) 2020 for higher education is making way for large-scale, transformational reforms, where the main focus is innovative research with multidisciplinary insights. The formative assessments with continuous class evaluation will motivate the students by increasing their participation and class attendance. Learning with peers has more impact with an ease, while understanding various difficult concepts. Group discussion activities; make it a fun for learning with active participation of the students. Apart from physical classrooms, the online classroom, such as Google Classroom, also acts as a good teaching platform. Students discuss their queries, lab difficulties or problem-solving difficulties in these discussion forums.

2.4 LEARNER-CENTRIC CLASSROOM TEACHING

The classroom teaching, when merged with fun and peer-learning activity, it motivates the interest of learning in the learners. It is observed that due to shy nature of many students, they like peer learning and try to open up with their friends. The flipped classroom concept inculcates the self-learning approach among the students. Some of the good and technical videos are shared with students. The students are asked to study those videos in a week's time. Then in classroom, groups of four members are formed and problem sheets are circulated among them. The facilitator interacts with students in a group or one to one, while solving the queries of the students. Hardly any concept needs to be revised in the class. It's observed that there's 100% involvement of the class and complete understanding of the concept is achieved through Flipped Classroom activity. Different experiences of second-year B.Tech students have been discussed with active learning activities in their classes of various courses. Also, the analysis of the active learning processes adopted was done with effective outcome.

As shown in Figure 2.2, the active learning session interactions of students among themselves and their peers make the process of learning more simple and easy to grasp. The group learning in classroom creates enthusiasm in the students to learn a new concept. Figures 2.2 and 2.3 are the real-time charts of the classroom session activities conducted for second-year students of Engineering. The task helped the students to learn new programming language and its salient features by exploring their existing knowledge. The students could comprehend the difficult tasks by communicating with their colleagues. Also, explaining the concept with the real-time day-to-day examples assist the students to grasp the idea in a better way. Today, the *Smart Classrooms* are the need of future to engage the students with creative learning. The interactive audio–video teaching aids are more expressive than traditional blackboard teaching. Thus, the new digitized *Learner-centric Era* with *Smart Teaching* aids is the solution to the traditional classroom teaching. The innovative classroom teaching practices will improve the teaching outcome with more realistic learning.

FIGURE 2.2 Active learning session discussions.

FIGURE 2.3 Innovative classroom teaching: Learning the features of a programming language.

2.5 LITERATURE REVIEW

The literature study reveals that Wenche Elisabeth Thomassena, Elaine Muntheb has done qualitative study about Norwegian pre-service Teachers (PSTs) orientations toward teaching in multilingual and multicultural classrooms by contributing to global research in this field. April Salerno et al. have explained how the secondary pre-service Teachers (PSTs) perceive content-area literacy integration through practitioner-research study of how the PSTs constructed Quad Text Sets (QTS). Results indicated that PSTs envisioned QTS as a means to teach content-specific background knowledge and to a lesser extent content-specific literacy and general literacy. Jeffrey et al. have created and disseminated a survey regarding educators' Instagram use. Analyses of 841 responses suggested that participants are engaged in the exchange of both professional knowledge and its real-time application. Authors have also discussed the implications of these findings for educators' work in the digital era and the future of research on educators' social media activities. Solange et al. have investigated differences in emotion recognition among schoolchildren experiencing the Montessori versus traditional practices. Children performed two tasks: (1) measuring the impact of social context on fear-surprise perception and (2) measuring their bias toward happiness or anger. Results suggest that children experiencing traditional practices show a higher sensitivity to fear-recognition, while children attending Montessori schools show a higher integration of social cues and perceive expressions of happiness for longer durations. Three experiments were conducted to explore whether adding eye movement modeling examples (EMME) to a short-narrated animation could facilitate visual processing during multimedia learning and learning outcomes, and whether the effects of EMME depended on the pace of lesson or prior knowledge level of the learners. In conclusion, adding visual signaling (or cueing) in the form of EMME can guide visual processing and improve learning outcomes in multimedia learning with a short lesson.

Problem-based Learning (PBL) is an instructional approach and their characteristics have been used successfully in multiple disciplines. John Savery explains the method that endows learners to conduct research, integrate theory and practice and apply knowledge and skills to develop a viable solution to a defined problem. It also explains similarities and differences between PBL and other experiential approaches like Project-based Learning, Case-based Learning and Inquiry-based Learning. Using this method, students will be able to access lots of information that was unheard-of a decade ago and find out more than enough problems to choose from in a range of disciplines. In author's opinion, it is vitally important that current and future generations of students experience a PBL approach and engage in constructive solution-seeking activities. As enrollments in online courses continue to increase, there is a need to understand how students can best apply self-regulated learning strategies to achieve academic success within the online environment. Broadbent et al. and other researchers focus on the different strategies of time management, meta cognition, effort regulation and critical thinking, which are positively correlated with academic outcomes, whereas rehearsal, elaboration and organization have the least empirical support. The authors also suggested that peer learning had a moderate positive effect; however, its confidence intervals crossed zero. The traditional lecturing learning is compared with active learning in undergraduate Science, Technology, Engineering and Mathematics (STEM) courses. Scott Freemana et al. talk about the innovative practice of active learning technique; where it was observed that on an average, students' performance increased by 0.47 % over the traditional lecture method for 158 studies. Also failing percentage was reduced to 1.95 for 67 studies. This heterogeneity analyses indicated that both results hold across the STEM disciplines that active learning increases scores on concept inventories more than on course examinations and that active learning appears effective across all class sizes—although the greatest effects are in small ($n \leq 50$) classes. The research also discusses that Trim & Fill analyses and fail-safe in calculations suggest that the results are not due to publication bias. The results also appear robust to variation in the methodological rigor of the included studies, based on the quality of controls over student quality and instructor identity. Author claims that this is the largest and most comprehensive meta-analysis of undergraduate STEM education published to date. The results raise questions about the continued use of traditional lecturing as a control in research studies and support active learning as the preferred, empirically validated teaching practice in regular classrooms.

2.6 21ST-CENTURY COMPETENCIES

Learning in collaboration is one of the effective learning method used in 21st century. Learning in teams, through group discussion and problem-solving by helping each other, has become a trend in 21st century. Marjan Laal et al. have used experimented problem-based team learning among the engineering students and observed that it is very useful for all types of learning. In collaborative learning, learners get benefits of academics, social and psychological sharing. Traditional learning methods are not effective due to lack of immediate feedback. Currently due to technological advancement, it is possible to check and assess learners' understanding and performance. As per Simelanea Andile Mji et al., the Technology engagement Teaching strategy is very much useful for effective learning. Active learning can be achieved with the use of different technologies like Google Classroom, Learning Management System (LMS) like Moodle, Microsoft Teams, Google Meet, etc. Formative assessment can be conducted using platforms such as Canvas, Kahoot, Mentimeter, etc. These educational platforms make learning a fun with proper conceptual understanding. We, the authors, too are using said technologies at our university on regular basis and found it effective. In 21st century, learners get all required information on internet on a single click. So teachers are facing problem of keeping students attentive in a class. Active learning methods like flipped classroom, Think Pair Share or problem-based learning methods are useful to overcome the said problem. Teachers can make use of technical advancements to keep learners active.

2.7 SMART CLASSROOM PRACTICES

The Smart Classroom practices can help to covert the traditional teaching–learning processes into a Smart Teaching Pedagogy with interactive student involvements, giving a solution with various forums and research discussions. Before using active learning methods, it was observed that the active involvement of the students was less. Also the attendance in class was low. The teaching was found to be one way only. Thus the self-learning with interest was lagging somewhere among the students. The problem-solving skills of the students were poor without any creativity and research.

The literature survey for such problems was studied to find feasible solution and outcomes. The active learning methods were used for the second-year Engineering students with courses such as "Principles of Programming Languages", "Data Structure", etc. The innovative teaching pedagogies were adapted by the faculties. Now, the modified classroom teaching sessions comprised of various modules like Flipped classroom, Google Classroom, Think-Pair-Share (TPS) activity, PBL, etc. As shown in Figure 2.4, the active learning pedagogies will motivate the students and increase their active participation in two-way teaching process.

Some of the active learning methods are discussed below.

2.7.1 Google Classroom

Google provides a facility of creating virtual classroom of the teachers. Using a Google Classroom, we can create discussion forum of students, post questions on the forum and ask the students to reply with their understandings. The facilitator can observe the responses of the students on forum and use it for evaluation purpose. This also can be used as one of the graded activity.

2.7.2 Flipped Classroom

Flipped Classroom changes the traditional teacher's role to a Facilitator. Teacher will not teach the concept, but the students will learn it themselves. Students will learn

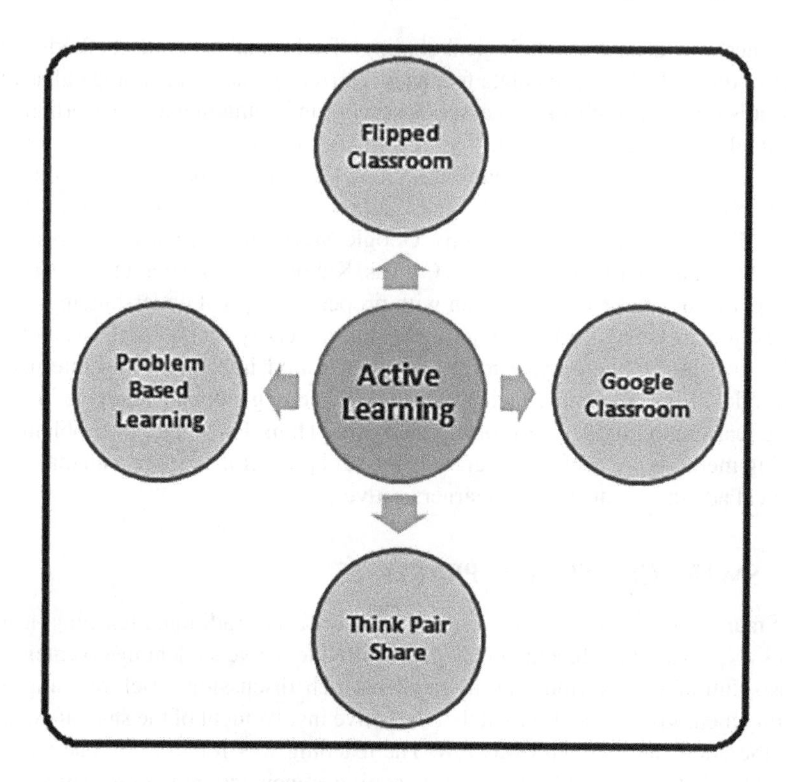

FIGURE 2.4 Active Learning Methodology.

outside the class and teacher will do the task of Facilitator inside the class. For implementing flipped classroom module, following are the steps:

- Good videos with duration (10–15mins) are searched and selected for the topic for which flipped activity is to be implemented.
- Problem Statements are designed on respective topic.
- Students are asked to watch videos.
- In the class, problem statements are distributed among students group.
- Students are asked to discuss and solve problems in group.
- The Facilitator moves around in the session and solves queries of the students.
- Rubric is used to evaluate the students, for example, as shown in Table 2.1.

2.7.3 Think-Pair-Share (TPS)

TPS is one of the effective active learning methods, as the students have the comfort for discussion with their peers. This activity is conducted on the concept, which is already taught in the class. Below steps are taken in this method.

- Pairing the students.
- Designing problem statements.
- Distributing problem statements among the students.
- Initially the students are asked to solve problems individually.

TABLE 2.1
Assessment with Rubric for Flipped Classroom

Not watched video, understanding not clear, with teacher's help also not solved problems	Watched video, understanding not clear, but with teacher's help solved problems	Watched video, understanding is clear, but not able to solve problem	Watched video, understanding is clear and problems solved with teacher's help	Watched video, understanding is clear and able to solve problems on their own
0	2	4	6	8

- Solution will be discussed and compared among the paired partner.
- Finally the Facilitator will verify the solution of all students.

As a part of active learning activity for DS-I subject, the Think-Pair-Share activity was conducted in the classroom on the topics such as Array-Strings, Searching-Sorting and Sparse-Matrix. A group activity was conducted in the lecture with duration of 45 minutes to 60 minutes and the groups are formed with at most four students. Some of the outcomes expected at the end of activity were like problem-solving, tackling difficult problems of similar kind, working in group/team, etc.

There were 13 groups with 63 students. Two Facilitators conducted the activity and the performance of the students was analyzed as seen in Figure 2.5. The analysis

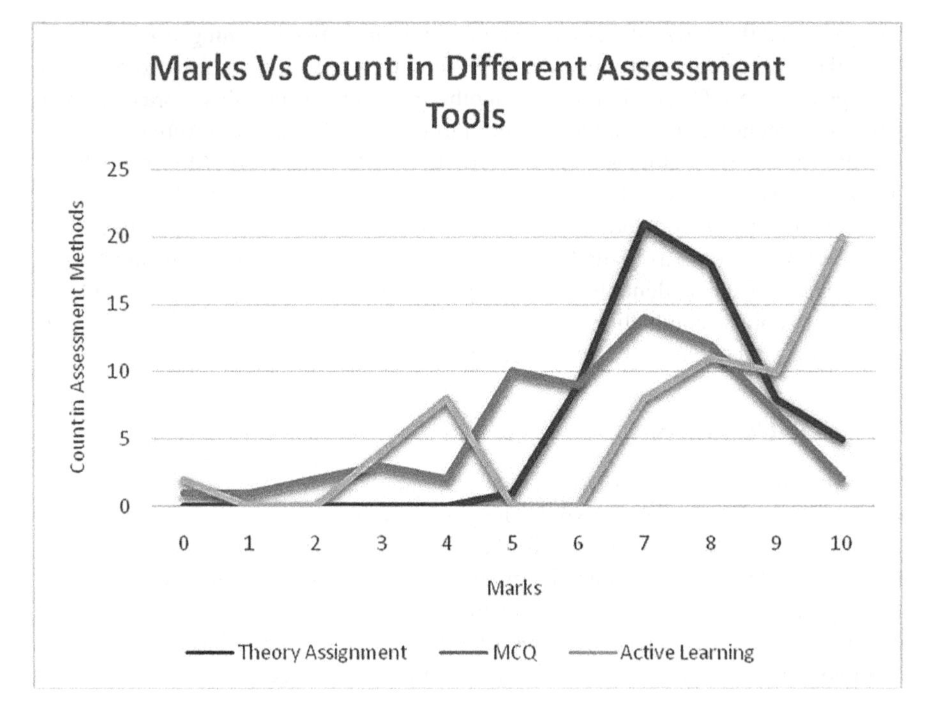

FIGURE 2.5 Analysis of implementing AL activities.

TABLE 2.2
Active Learning Techniques

Sr. No.	Active Learning (AL) Method	Parameters			Advantages of AL
		Topic	*Duration*	*Nature*	
1	Google Classroom	Discussed in class	2–3 Lectures	A complete outside the class activity	✓ Active participation of students with more enthusiasm ✓ Self-learning was motivated ✓ Attendance in class was more ✓ Some students who were shy began to open their minds while learning with peers. ✓ Problem-solving skills were improved ✓ Team skills were enhanced ✓ Overall performance of the students improves
2	Flipped Classroom	Self-learning	1 Lecture	Semi class activity	
3	Think Pair Share (TPS)	Discussed in class	1 Lecture	Full class activity	
4	Problem-based Learning (PBL)	Discussed in class	2–3 Lectures	Semi-class activity	

study revealed that the students scored more marks in Active Learning (AL) activities then other two assessment methods. TPS Activity helped the students to learn the concept very well. Discussion with each other makes their thoughts more clear with better understanding. 65% students score more than 80% marks in Active Learning activity—TPS, 33% students score more than 80% marks in Multiple Choice Questions (MCQ) and 49% students score more than 80% marks in Theory Assignment. Thus it concludes that Active Learning Activity—TPS gives the excellent result while comparing the MCQ and Theory Assignment tools. The analysis of performance of the students for the active learning activities was assessed with the continuous course conduction. Table 2.2 reflects the impact of implementing the active learning techniques for the second-year students of Engineering.

2.8 CONCLUSION

A person knowing design principles with its implementation and research activities will survive and serve the mankind. The youths are not only the building blocks of our Nation but the whole Universe. To shape these youths and future of tomorrow, we have to provide more experience learning with active participation of students. The learning styles of teacher and student may be different but what is important is acquiring knowledge. Science and Spirituality together will make it easy to grasp the technology with the peace of mind. Thus, it's the modern pedagogy that has learner diversity, where the Teacher has to cater to the diversity of learning styles and acts more like a Facilitator. The active learning pedagogies will definitely motivate and increase the active participation of the students in classroom.

BIBLIOGRAPHY

Bossér, U., Lundin, M., Lindahl, M., & Linder, C. (2015). Challenges faced by teachers implementing socio-scientific issues as core elements in their classroom practices. *European Journal of Science and Mathematics Education, 3*(2), 159–176.

Broadbent, J., & Poon, W. L. (2015). *Self-regulated learning strategies & academic achievement in online higher education learning environments: A systematic review*, The Internet and Higher Education, Elsevier.

Carpenter, J. P., Morrison, S. A., Craft, M., & Lee, M. (2020). How and why are educators using Instagram?. *Teaching and teacher education, 96*, doi:10.1016/j.tate.2020.103149

Darling-Hammond, L., Flook, L., Cook-Harvey, C., Barron, B., & Osher, D. (2020). Implications for educational practice of the science of learning and development. *Applied Developmental Science, 24* (2), 97–140. doi:10.1080/10888691.2018.1537791

Denervaud, S., Mumenthaler, C., Gentaz, E., & Sander, D. (2020). Emotion recognition development: Preliminary evidence for an effect of school pedagogical practices. *Learning and Instruction, 69*, doi:10.1016/j.learninstruc.2020.101353

Edström, K. (2018) Academic and professional values in engineering education: Engaging with history to explore a persistent tension. *Engineering Studies, 10*(1) 38–65.

Foncha, J. W., Abongdia, J. F. A., & Adu, E. O. (2017, 25 Sep). Challenges encountered by student teachers in teaching English language during teaching practice in East London, South Africa. *International Journal of Educational Sciences*. doi:10.1080/09751122.2015.11890302

Freeman, S., Eddy, S. L., McDonough, M., Smith, M. K., Okoroafor, N., Jordt, H., & Wenderoth, M. P. (2014). Active learning increases student performance in science, engineering, and mathematics. *Proceedings of the National Academy of Sciences, 111*(23), 8410–8415, www.pnas.org/cgi/doi/10.1073/pnas.1319030111

Green, K. (2010, July). 5 Basic elements of cooperative learning, Ontario; Canada. Retrieved 20 Dec. 2011, http://cooperativelearning.nuvvo.com/lesson/216-5-basic-elements-of-cooperative-learning.

Green, S. K., Johnson, R. L., Kim, D. H., & Pope, N. S. (2007) Ethics in classroom assessment practices: Issues and attitudes. *Teaching and Teacher Education, 23*(7), 999–1011. ISSN 0742-051X, doi:10.1016/j.tate.2006.04.042.

Grunert, M. (2009). *Cohen, the course syllabus: A learning-centered approach*, ISBN-13: 978-0470197615, John Wiley & Sons.

Hake, R. R. (1998). Interactive-engagement vs. traditional methods: A six-thousand-student survey of mechanics test data for introductory physics courses. *American Journal of Physics, 66*, 64–74.

Harden, R. M. (2009a). Developments in outcome-based education. *Taylor & Francis Online*, 117–120, doi:10.1080/01421590220120669.

Harden, R. M. (2009b). Outcome-based education: The future is today. *Taylor & Francis Online*, 625–629, doi:10.1080/01421590701729930.

Hsu, A., & Malkin, F. (2011). Shifting the focus from teaching to learning: Rethinking the role of the teacher educator. *Journal of Contemporary Issues in Education Research, 4*(12), 43–50.

Johnson D.W., & Johnson R.T. (2009). An educational psychology success story: Social interdependence theory and cooperative learning. *Journal of Educational Researcher, 38*(5), 365–379.

Laal, M., Laal, M., & Kermanshahi, Z. K. (2012). 21st century learning; learning in collaboration. *Procedia—Social and Behavioural Sciences*, *47*, 1696–1701.

National Education Policy. (2020). *Ministry of human resource development*. Government of India.

Panitz, T. (1996). *A definition of collaborative vs cooperative learning*. Deliberations, London Metropolitan University; UK., Retrieved 20 Dec. 2011, from: http://www.londonmet.ac.uk/deliberations/collaborative-learning/panitz-paper.cfm

Roehl, A., Reddy, S. L., Shannon, G. J. (2013). *The flipped classroom as a way of engaging the Millenial through active learning strategies*. Department of Interior Design & Merchandising at Texas Christian University.

Salerno, A. S., Brown, A., Rutt, A., & Heny, N. A. (2020). How pre-service teachers construct quad text sets for use in secondary classrooms. *Teaching and Teacher Education*, *96*, doi:10.1016/j.tate.2020.103147

Savery, J. R. (2006). Overview of problem-based learning: Definitions and distinctions. *Interdisciplinary Journal of Problem-Based Learning*, *1*, (1), 3.

Simelan S., & Dimpe D. M. 2011). Clicker technology: The tool to promote active learning in the classroom. In R. Corchero (Ed.), *Education in a technological world: Communicating current and emerging research and technological efforts*, 83–98, Spain: Formatex Research Center

Simelane, S., & Mji, A. (2014). Impact of technology-engagement teaching strategy with the aid of clickers on student's learning style sibongile. *Procedia—Social and Behavioural Sciences*, *136*, (2014), 511–521.

Simelane, S., Mji, A., & Mwambakana J., (2011). *Clicker-technology teaching strategy and students approaches to learning in synchronized activities. Proceedings of the World Conference on E-Learning in Corporate, Government, Healthcare, and Higher Education 2011*, Honolulu, AACE, 1708–1713).

Simelane, S., & Skhosana, P. (2012). Impact of clicker technology in a mathematics course. *Knowledge Management & e-Learning: An International Journal (KM& EL)*, *4*(3), 279–292. Retrieved from http://www.kmel-journal.org/ojs/index.php/online-publication.

Srinivas, H. (2011 Oct 21). What is Collaborative Learning?, The Global Development Research Center, Kobe; Japan, Retrieved 20 Dec. 2011, from: http://www.gdrc.org/kmgmt/c-learn/index.html

Thomassen, W. E., & Munthe, E. (2020). A Norwegian perspective: Student teachers' orientations towards cultural and linguistic diversity in schools. *Teaching and Teacher Education*, *96*, 1–11, doi:10.1016/j.tate.2020.103151.

Walker, S. E. (2003). Active learning strategies to promote critical thinking. *Journal of Athletic Training*, *38*(3) 263–267.

Wang, F., Zhao, T., Mayer, R. E., & Wang, Y. (2020). Guiding the learner's cognitive processing of a narrated animation. *Learning and Instruction*, *69*, doi:10.1016/j.learninstruc.2020.101357

3 Leveraging Information and Communication Technology for Higher Education amidst the COVID-19 Pandemic

Zahid Hussain Bhat
AAA Memorial Degree College,
Cluster University Srinagar, India

CONTENTS

3.1 INTRODUCTION

Coronavirus (Covid-19) began to spread exponentially worldwide by the end of 2019, according to a UNESCO survey, causing the death of over 3,000 individuals. Subsequently, many nations began to initiate related policies to combat this disaster together with the closure of schools. As on March 2020, 46 countries in five distinct continents declared closures of schools and colleges to curb the dissemination of this pandemic (CoSN, 2020). With the passage of time, millions of students were unable to attend their colleges and universities due to nationwide lockdowns.

The Spanish-flu pandemic in the year 1918 caused the closing of schools worldwide, similar to the ongoing global coronavirus epidemic, and, generally, halted the formal schooling. A big contrast between these two epidemics is that there was no

DOI: 10.1201/9781003102298-3

prospect of bringing schooling online at the time; and however, even in the modern period, electronic learning has only been widely accessible and omnipresent in the last two decades with improved accessibility, connectivity and mobility of electronic devices. The ongoing pandemic triggered by Covid-19's worldwide dissemination is being identified as an unprecedented global epidemic.

Mentions of pandemics and future repercussions for academia are difficult to find in the literature on academic technologies, learning and communication. Indeed, in health journals and epidemiological journals, historical mentions of disruptions in learning due to the Spanish-flu pandemic of 1918 are scantly found (Stern et al., 2009). Parker White et al. (2010) noted the need to act in response to "warnings of impending pandemics" following SARS of 2003 and H1N1 of 2009. In education research, the present catastrophe appears to have been entirely unimagined in general. Although there is currently a lack of regular research on a nationwide level, the general trend is that teachers adapt "traditional" pedagogies in the classroom, develop videos and organize webinars and share materials and assignments online. What had been done head-on previously is now done online. More student-focused activities such as applying knowledge to tasks in practice, organizing peer reviews or making collective learning are now carried online. The development of all these events requires explicit instructional, technological and content knowledge as well as skills of the teachers. This trend seems to be confirmed by the research carried on the experiences of online training of faculty by higher learning institution along with social media and TV.

A redefinition of the success of the numerous educational agents has included the convergence of the use of technology in classrooms and the creation of new methodologies for the forthcoming horizons in order to predict the current technical developments that will have an impact in the coming years. The definition of technology as the axiom of the environment and culture today, as well as its relevance in educational contexts, has been illustrated by the emergence of various research and studies in which emerging technologies in education have been the key subjects, especially during the last decade. The constant and rapid development of technology has made the academic and science world concentrate on studying the advantages and drawbacks of the use of Information and Communication Technology (ICT) in classrooms, especially in the field of education.

Using the basic axes of usability, equanimity and equality, along with consistency and all learning outcomes based on competency-based learning, UNESCO (2015) has decided that the prospect of ICTs needs to be used to achieve its targets by 2030. It highlights mobile learning, open educational resources (OER), the need to have access to open-access publishing (particularly for professionals) and open-source applications. Massive Online Open Courses (MOOCs) have now been one of the main aspects of permanent and modified training with respect to lifelong learning. At the start of the 21st century, the role of the internet and digital technologies in higher education was unquestionable. In this sector, the advancement of technology causes rapid changes. Participants of the school system continue to face new obstacles because of the globalized world and advances in ICTs. Fresh questions and topics are arising that need to be answered. By building new technologies into its work, higher education is seeking to keep up with technical growth.

As a result, governments have been encouraged to improve their pedagogical efficacy and to offer certifications that are accredited. The emergency of the produced data in the network is not ignored by UNESCO; that is why it has developed learning analytics and data mining as a viable environment that forges the path to a better understanding of the learning process of the students, as well as the ability to adapt and enhance the environments listed in the learning processes, particularly those performed online and always respecting the individuals' protection and privacy. Finally, it promotes the value of accrediting and validating all learned expertise, skills and competencies from informal, formal and non-formal educational contexts. In fact, in a higher education context, several studies are being performed that aim to address this field via blockchains.

3.2 OVERVIEW

Technological advances impact social change by influencing and modifying daily contact habits, the acquisition of knowledge, free time and learning. The process of learning to discover new forms and mechanisms is necessary in a knowledge-based society. The cultural, motivational, communication and social disparities, perceptions and demands of students in higher education are becoming more evident as the social and economic conditions of change. Changes because of technical advances are difficult to adopt and in the meantime, ideas are evolving in relation to research and knowledge retrieval. Higher education has been influenced and the academic environment has been fundamentally changed by the exponential growth of ICT. This indicates the presence of new opportunities and challenges and a positive impact on the growth of e-learning. E-learning is an educational practice that uses electrical technologies or resources "(Visvizi, Lytras, & Sarirete, 2019), including computer learning or the possibilities of an online educational network." According to the wildest theory, e-learning is every learning act that takes place through the use of information and communication resources." It helps to develop the knowledge market by providing new and additional answers to training requirement. E-learning is every learning act that takes place by the use of information and communication tools.

Several guidelines were released to include the spread of the Covid-19 pandemic by UNESCO to ensure continuous teaching and learning that emphasizes preparation, data protection, curriculum inclusion, psychosocial difficulties, course schedule, instructor and parent support, the limit of the number of platforms/applications, student learning process control, specifying the length of distance learning. In this moment of chaos and confusion brought upon us by this pandemic, the future of various learners has tumbled. Most students now live under drastically different conditions and are required to uphold their academic performance than before the withdrawal of head-on classes in universities. For certain learners, the head-on lecture hall and its related events have been the most familiar place of their radically altered lives, amid these obstacles. Large-scale national attempts to leverage technologies to ensure remote and online learning have been used to mitigate the consequences of the Covid-19 pandemic (Miks & McIlwaine, 2020). More recently, top universities have turned to creative modes of teaching and learning, such as "MOOCs" (Brahimi & Sarirete, 2015; Brahimi, Khalifa & Benaouda, 2019; Visvizi, Lytras, & Sarirete, 2019), which have drawn not only the attention of higher education educators but also the interest of high schools. The college classroom is a changing constant throughout this pandemic. In addition, the link of students with their class is their only option to normality, and the class is their only opportunity to reach beyond their families during the lockdown. Students are asking for certain teachers to resolve their life conditions and the difficulties they face, as faculty and students both altogether are seeking to meet their scholastic objectives. While the transfer to online teaching and learning can sound like a provisional transition needed by this global epidemic, and both faculty and learners are prospective of predicting that situation will move back to normalcy in future, this study argues that the demand for the pandemic presents both faculty and students with a vital and special learning opportunity. This transition was uninvited and hence unintentional for most faculty and students, if not most. In certain situations, in the unexpected conversion from head-on environments to simulated classes, both parties became unwelcome participants. The new simulated classroom has become the eventual gathering ground for educational sessions, whereby teachers and students have to work out how to reap the benefits of an extraordinary situation in real time. The basic understanding is that these online curriculum arrangements offer a remarkable prospect for faculty to re-imagine their instruction and to develop some very significant instructional and technical skills for our students that they may not have taken into account so far.

Imparting professional education in a pandemic can be used as mechanism for creating stability by giving both students and teachers a sense of normalcy and intent in responsive and substantive ways. This need for normalcy and stability is a latent desire among students that many faculty in their simulated classes have already encountered. Furthermore, through the interactive classroom, faculty are engaged with designing plans and methods to address these needs. It is fairly clear that a growing number of job environments have become distant and will continue to be so, when operating from home was implemented as a reaction to the pandemic. Approaches to educate learners in good behavior and ethical conduct are also skills that can be both useful and transferable in potential job environments while in the online virtual classroom. In addition, other skills needed within the virtual space for efficient learning are often transferable to prospective jobs settings.

3.3 INDIAN CONTEXT

In India, the different restrictions and the national lockdown for Covid-19 has affected more than 32 crores of students. In an attempt to control the spread of the Covid-19 pandemic, most governments around the world have temporarily shut down educational institutions. The world's student population has been drastically affected by this global closure. Governments around the world are making efforts to mitigate the immediate impact of the closure of educational institutions, especially for more vulnerable and disadvantaged communities, and are seeking to facilitate the continuity of education through various digital learning modes for all. According to a survey report conducted on higher education by the Government of India's Ministry of Human Resource Development (MHRD), it was noted that there are 993 universities,

39,931 colleges and 10,725 standalone institutions listed on their portal that contribute to education (Visvizi, Lytras, & Sarirete, 2019). Although the country has adapted to new-age learning, there is still an obstacle to achieving full success because only 45 crore of the country's total population have access to the internet and e-learning. People living in rural areas are now seriously deprived of infrastructure, hampering the root cause of online education. By allowing educational institutions to promote online learning and adopt a virtual learning community, the Covid-19 pandemic educated the entire world about how necessity is the mother of innovation. With technological innovation and progress, the pandemic has steered the education sector forward. This global epidemic has disturbed higher education sector significantly. A considerable number of native scholars registered in several foreign universities are now leaving those countries especially in the worst affected countries, and if the situation continues, there will also be a substantial drop in call for foreign higher education in the long run.

During this pandemic, the institutions of higher education have responded positively and adopted different strategies to face the crisis. A number of preventive measures were also taken by the Government of India to prevent the spread of the Covid-19 pandemic. By launching many virtual platforms with online depositories, e-books and other online teaching/learning materials, channels of education through DTH, radio stations, the MHRD and University Grants Commission (UGC) have made several arrangements. Students use common social media platforms such as WhatsApp, Google Meet, Zoom, Facebook Live, YouTube Live, Telegram, etc. during lockdown for online method of teaching. MHRD's (https://mhrd.gov.in/ict-initiatives) ICT initiative is also a matchless policy that brings together all online education digital resources (Pravat, 2020). In view of the Covid-l9 pandemic and subsequent lockdown on 29th April, 2020 (UGC notice), UGC has issued Guidelines on Examinations and Academic Calendar. All terminal examinations have been postponed and moved to November 2020 and classes have been proposed to commence next year. In view of the lockdown, UGC has also prepared a full schedule for the 2020–2021 academic session with new dates.

The pandemic of Covid-19 has shown a huge gap among our students in the digital divide. A student would need to have access to the platform and parents' guidance in order to learn online via these platforms. In terms of access, having the internet and a device such as a computer, laptop, tablet or smartphone would be a prerequisite for learning via an online platform. There are complaints in urban areas that kids have to share devices, and because of slow internet connections or lack of devices, they could not fully use the lessons online. Parents would need to be educated in order to assist and provide the children with guidance. Children with less educated parents would be disadvantaged and left to figure out how to use and learn via the platforms themselves. In terms of digital preparedness, this crisis has also shown that Indian education system is lagging behind. Most countries worldwide that are trying to transfer learning from classrooms to online mode are concerned with this problem. Many educators are not technology-savvy and are forced to adapt to online mode to conduct their classes. Such online platforms have always been a supplementary teaching material for classroom use. Now, it has become the only way to reach and educate learners.

3.4 THE RATIONALE

Many nations have been influenced by Covid-19 pandemic in the evolving and transforming situation. Similarly, most schools, colleges and universities have been closed and institutions are switching quickly to new ways of teaching and learning. The educational system has been profoundly impacted by the Covid-19 pandemic. Due to the resultant sense of confusion and distress among students and faculty members, there is a good risk of worsening mental health (Sahu, 2020). Students are facing significant interruptions in studying and teaching in these circumstances; graduations could also be postponed, not to mention the associated economic difficulties (Moopen, 2020). Online learning has been increasingly innovated and adopted by organizations, attributable, in part, to proven familiarity with the requisite resources, teaching methods and online learning considerations. For certain students reluctant to return to in-person courses, this has resulted in less disturbance.

Undoubtedly, institutions which do not have such training and preparation now need to set their affairs right to prevent the unnecessary burden and conflicts which will emerge from rapidly implementing ideas. In this crisis, there is need for online learning skills, and it should act as a reminder that organizations must develop this capability as quickly as possible. Across the globe, policymakers are searching at steps to safeguard their students, while exploring opportunities to carry on offering online classes. Literature indicates that school authorities advise workers in the quarantined and shut down areas to collaborate and exchange knowledge and digital infrastructures for online instruction (Czerniewicz, 2020; Visvizi, Lytras, & Sarirete, 2019). In order to continue teaching, countries most impacted by the global epidemic are trying their part. For example, before the whole world was quarantined, Italy was the first country to shut its educational institutions and switch to virtual classes. Similarly, other countries followed the suit and adjourned all head-on instruction. Literature shows that educational institutions are taking charge internationally and are doing all that they can to slacken the Corona virus's further dissemination. This reform is necessary, according to Czerniewicz (2020), as it is needed by the existing pandemic situation for safeguarding faculty and students. In keeping with this issue, organizations around the world saw the need to transition to teaching and learning using ICT.

In addition, the rapid development of ICT and the growing uncertainty that comes with its exploding capacity explains why, particularly in the aftermath of the Covid-19 pandemic, technology convergence in education continues to receive particular attention. The beginning of the new millennium witnessed the onset of the internet and its initial entrance into our educational institutions, which demanded faculty to prepare themselves for a different line of learners possessing unique desires and characteristics. This is due to the fact that their arrival into this world coincided with technological developments throughout the world which was pervasive and universally embraced. This study may help to map and understand the usage of digital tools and behavior in online environment of different generations in higher education. Therefore the research can be of help to higher education in finding its aims, regarding the development and delivery of curriculum supported by new technologies.

3.5 INTEGRATION OF ICT IN HIGHER EDUCATION

ICT has become an important part of our daily life and has changed the academic atmosphere to the degree that knowledge of ICT for almost all qualifications has become a practical necessity. In addition to changing the way students learn, the introduction of technology in education has also transformed the teaching methods by encouraging interactive practices (Haddad, 2003). Both conventional obstacles have been removed by the developments of the internet and the World Wide Web and the idea of information exchange has been further reshaped (Michelsen & Wells, 2017; Visvizi, Lytras, & Sarirete, 2019). In addition to offering rich instructional tools, contemporary online learning platforms have the ability to facilitate real-time and asynchronous collaboration between teachers and learners. This is in spite of the fact that for decades, online learning has been looked down upon in higher education (Shachar, & Neumann, 2010). Online learning environments facilitate additional learning opportunities in which students can chat, interact and take charge of their individual learning at their own speed and time. As a result, ICT engrossed programs offer our students with an inspiring and stimulating learning experience and often contribute to their self-directed learning. The position of the instructor, at the time when the transition from a teacher-administered atmosphere to a student-administered atmosphere happens, becomes more of a facilitator and as such minimum support might be needed (Geng et al., 2019). The teachers are clearly the core players in the successful application of ICT blended learning in the form of instructors, teachers and professors (Sipilä, 2011; Aydin, 2012; Buabeng-Andoh, 2015). It is, therefore, vital that they possess the right kind of mindset and opinions about ICT for successful incorporation of technology in education. The opinions and expectations of students must also be taken into account as it specifically affects their liberty and their styles of learning (Jung, 2005; Fu, 2013; Buabeng-Andoh & Totimeh, 2012; Mirzajani et al., 2016).

It must be identified that adequate technological support in the form of infrastructure, tools, hardware and software support systems be provided for productive online and blended learning. This is for certain that the inclusion of ICT as an educational tool in undergraduate programs has grown exponentially. Universities and colleges have consequently begun to introduce software and instructional blogs to complement current education and practice (Ruzgar, 2005; Becker, 2000). Meta-synthesis of related literature indicates that, through the use of technology to increase the eminence of education and learning, there has been a mounting appeal in the creation and usage of multimedia enriched e-content in recent years (World Bank, 2020; UNESCO, 2020; Smith & Judd, 2020; CoSN, 2020). Visual educational materials incorporating text, images, audio and animations are interactive content. Teachers prefer to employ these resources to help, show and clarify complicated topics that cannot be readily illustrated using text alone for enhanced learning in the classroom (Thomas & Israel, 2013; Lanzilotti et al., 2006). Studies have demonstrated that many advantages are offered by the effective use of multimedia-enhanced content in an instructional setting. The use of ICT resources will help enhance the interpretation of instructional materials by learners (Lanzilotti et al., 2006). Learners can also simulate actual processes and can conduct simulated experiments which otherwise would be risky and

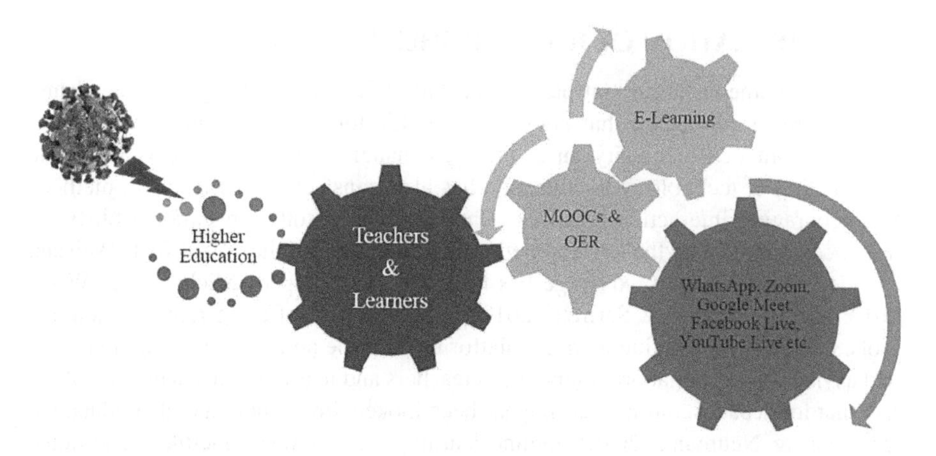

FIGURE 3.1 COVID 19 Impact and Leverage of ICT.

costly to perform in a laboratory (Hennessy et al., 2006). In particular, the speed of technology iteration must be increased and the technological implementation of online education programs streamlined, taking into account the troubling impact of Covid-19. In continuing teaching and research programs while adhering to the extended closures, this pandemic has presented a threat to the learning institutions. Consequently, in these extraordinary times, this chapter plans to study the integration of ICT in higher learning institutions (Figure 3.1).

Despite some shortcomings in the organization and execution of online learning, the extraordinary state of affairs in the aftermath of this pandemic has created challenges for faculty, students and administrators to accept online mode of learning. Literature highlights those shortcomings, e.g. the vulnerability of the online learning structure, the knowledge deficit, inexperience of students and dynamic home climate (Murgatrotd, 2020). In spite of such restrictions, however, the present scenario needs intervention such that the student's schooling is in no way compromised. On the backdrop of the Covid-19 pandemic governments and tertiary institutions around the world are launching numerous policy programs to continue teaching practices in order to suppress the virus and benefit the student community. There is no doubt that the world's massive technological developments need an exemplar change in the way we accomplish our academic aims and ambitions. During the pandemic, large-scale global attempts are emerging and developing rapidly to employ technology for distant learning, online learning and distance education. ICT embedded learning using laptops, projectors, tablets, mobile phones and digital white boards have been introduced by numerous colleges and educational institutions. In addition, teachers have discovered that students share a unique connection with ICT. Children of today are globally introduced to electronic devices such as tablets and smart phones from their tender ages (Naciri et al., 2020). An observational research conducted by Jesse (2015) supports the above-mentioned observation. The results of his study indicate that most students use these devices (almost 99.8%) for chatting, texting, accessing social media and playing games. Considering the pace of introduction of technology into the education structure, it can be concluded that the level of adoption and receptivity of students toward it is high (Willms & Corbett, 2003). Teachers also claim that the students are very addicted to smart devices and are looking ahead to integrate technology into learning. The findings of Earle (2002) and Buabeng-Andoh & Yidana (2014) posit that learners are moving ahead with integrating ICT as it supports their education and learning using internet from the coziness of their homes. This motivation in learners calls for teachers and authorities to make available sufficient technology underpinning and learner support facilities for a satisfying and an eloquent learning. In the heart of it, it can be inferred that teachers are essential players in the efficient execution of a structured ICT learning and must be respected and supported accordingly. Nonetheless, the receptiveness of learners to ICT is well understood by both.

Education administrators understand that technology is certainly not impartial, and there is some degree of opposition and disagreement to any transition. Present political interests have attracted blended learning and extreme arguments have been made, such as one mode being stronger than another (Czerniewicz, 2020; Naciri et al., 2020). Meta-synthesis indicate that ability to accept transition is a vital requirement for effective technological implementation as it suggests opportunities for students to develop and relate the skills desirable in 21st century (Fullan, 2013; Ertmer & Otternbreit-Leftwich, 2019). In view of the persistent emergence of ICT in the field of education, its usefulness in optimizing the administration of lessons has been extensively debated and implemented internationally in many Higher Education Institutions (HEIs). This is for the reason that technology serves as a facilitator in the planning and execution of lessons and also assists faculty members (Sadegül Akbaba et al., 2011). One of the guidelines of the American Psychological Association has urged lecturers to consider introducing an ICT interactive learning atmosphere for students (Li, Yamaguchi, & Takada, 2018). For academicians, the Covid-19 pandemic is an essential adaptation and a transformative challenge, one for which there is no preconfigured blueprint that can lead toward appropriate responses. Leaders in education must design solutions efficiently and with specific circumstances in mind while the pandemic runs its course.

While in most countries, students having an access to internet and smart devices will not be the majority; however, assisting policymakers in the creation of successful modes of e-learning would open up institutional capabilities and resources to shift their attention on the provision of alternate learning approaches for learners who lack the similar prospects. The aim of this annotated collection of virtual academic tools for learners, faculty and parents is to help governments and other educational frontrunners in investigating and assessing various methods of continuing to educate students particularly during this pandemic. It can be used by those designing or improving upon a plan for education continuity, by either directly incorporating some of these resources into their plan or using them as a model to develop their own online educational materials.

Several revised principles of education underline that faculty members must show the ability to create learning environments and activities that incorporate technology successfully, particularly at the level of higher education. In addition, ICT is considered a significant agent for educational innovation and change, and there is a wealth of research that have shown that if properly applied, student success and quality of education can be increased (Elisha, Odini, & Ojiambo, 2013; Sadegül Akbaba, Kalayci, & Avci, 2011). In view of previous studies, evidence supports the view that skills in the use of ICT would be a crucial prerequisite for both faculty and students as time moves on (Coll, Mauri, & Onrubia, 2009; Naciri et al., 2020).

Universities have begun providing e-library facilities for 24 hours to facilitate the study of students and faculty members. There are collections of many journals, e-books and scholarly publications in these e-libraries (Solomon, 2020). Currently, many technological initiatives have been established and quickly extended to encourage distance learning, and internet for home learning, assisting countries in their attempts to provide continuing education by remote learning and to collaborate closely with education ministries in several countries to promote remote learning services related to coronavirus (UNESCO, 2020b). Teachers and educational institutions are now implementing technology-based methods in order to deliver instruction. For both teachers and students, there are several blackboard resources available. Home screen, Course and Instructional Resources, Updates, Chat Forums, Messages, Roster, My Groups, My Ratings (Grade Center), and the assessing tools for assessments, tasks, quizzes and tests. Platforms such as Google Connect, Zoom, Moodle, Google Classroom, Kahoot, Nearpod, Flipgrid, Slido, Quizizz and Google Teams, where users engage and connect and develop new skills, are also used in learning environments. Even one of the blackboard teaching resources is the computer classroom. Radio and national television programs are being used in many countries to ensure that educational lessons and resources are delivered, especially in under-resourced places that do not have access to technical infrastructure (ILO, 2020). To ensure inclusivity, effective and appropriate resources have been introduced for the provision of online learning. Increased access to technology has made it necessary to take major action to warrant data protection and safekeeping. Also, it was pertinent to discuss the psycho-social issues of students and parents during the quarantine, in addition to resolving instructional challenges. Students, instructors and parents must be trained in its use before carrying out the digital platform, and most notably, there must be certain methods to assess and track their learning in order to evaluate student learning (UNESCO, 2020b).

While the responses to addressing the influence of pandemic on higher education are no less inspiring and creative, the fact is that many colleges and universities are better equipped than others to take advantage of technical tools and facilities, helping them to adapt even more efficiently to educational emergencies during this pandemic. There is also a stronger need for educational institutions to improve instructional methods, and it would be of utmost significance to use novel teaching strategies and interventions (Toquero, 2020). In order to explain the use of mobile devices in educational environments, multiple reports on mobile learning have been performed (Sönmez et al., 2018; Naciri et al., 2020). Smart devices are now prevalent indeed. This potential indicates that it would be useful for both students and teachers to make use of mobile learning (Aubusson, Schuck, & Burden, 2009; Crompton, Burke, & Education, 2018). The benefits include mobility, i.e. the learning process is not confined to one unique location or any site, or any moment (Corbeil & Valdes-Corbeil, 2007). In addition, it enables teachers to configure teaching and enables learners to self-regulate their learning (Steel, 2012; Sha et al., 2012). Mobile learning typically helps students improve technical skills, conversational skills, find answers to their questions, develop a sense of teamwork, encourage the exchange of information and thus optimize their learning results (Al-Emran, Elsherif, & Shaalan, 2016).

Ali (2020) in a recent study posited that teachers in a tertiary organization appear to integrate ICT in their classes, but most of them (about 92%) still assume that trust is an aspect which needs to be improved (Ali, 2019). The management of the numerous instruments and learning platforms may be attributed to this lack of trust. Similarly, Huang & Liaw (2005) argue that the mindset of faculty members and their ability to adopt ICT make a significant difference in their students' lives. In view of the Covid-19 pandemic, the World Bank highlights the need for preparation and resources for faculty working online. They also advise that student learning will obviously not be assisted online by faculty who do not have access to ample internet and a wired computer at home (World Bank, 2020). However in reaction to the Covid-19 pandemic, the rush of universities to finish their online academic year could result in unparalleled challenges for teachers, students and families. This is because the way colleges and universities are actually integrating their online programs and courses may yield extremely uneven and unsatisfactory instructional learning experiences that can undermine the reputation of distance education as a viable and substantive educational medium (Durden, 2020).

The gradual transition to e-learning has been fruitful and it is possible to use the knowledge obtained in the future. A century of new rules, legislation, channels and strategies for potential cases will be compelled by the lesson learnt from Covid-19 (Basilaia & Kvavadze, 2020). E-learning is an integral component of instructional technology in higher education, considering the conditions of confinement created by the coronavirus and to ensure quality of teaching for students. It makes it easy to read, communicate and exchange ideas with students.

3.6 THE CHALLENGES

It is interesting to remember that to delay the propagation of the Covid-19 pandemic, the mass closing of universities has strongly pressured educators around the world to rapidly move their teaching and evaluations from face-to-face to online and distance learning. There was, however, a brief period for educators to train for this daunting phase of interactive teaching and learning. The incorporation of ICT in higher education is not immune to difficulties. Having a paradigm change in teaching and learning provides students and faculty around the globe with diverse challenges. The lack of studies on similar issues in the past, along with the confusion as to when new therapies and vaccinations for this pandemic will be available, has only brought an additional obstacle to the prediction of prospective HEIs' actions in the event of the suspension or restoration of face-to-face operations in the 2020–21 academic year.

3.6.1 Challenges for Students

From the point of view of the school, the challenges are that students had to leave the hostel and PGs immediately because of lockdown and they were unable to take their textbooks and laptops to their native and as such they did not engage in e-learning.

The first challenge is to test the access of the learner to internet communication. The students, in fact, come from rural and backward regions. To allow streaming of the live teaching, the students were supposed to have a smart phone with 4G compatibility. Except in urban areas, sufficient internet access to eliminate buffering and lagging of the live stream is also a problem. This will happen when the speed of the internet connection is insufficient. The World Bank also reiterates that it would be very difficult for most students to access online learning, especially those who remain with limited access to the internet and are subjected to various other difficulties (World Bank, 2020). The next challenge is that live broadcasting only relies on the theoretical distribution of subjects without the direct use of the classroom, because professional instruction is more realistic.

3.6.2 Challenges for Faculty

As second-line respondents, this plethora of student problems and desires lie under the unrecognized and unforeseen role of having to arbitrate an array of student needs and issues, further compounded by the indefinability of an emerging scenario. From the onset of lockdown in mid-March, the imminence of these issues has caused teachers to have constructive conversations with their students to explore the possibility of offering simulated online learning courses as per the prior schedule before this pandemic. Most students chose to attend these synchronous live virtual lectures in most situations, only few of them chose to provide asynchronous video classes to access course materials where they could. The larger obstacle for the faculty was also to have multimedia tutorials, to share room with members of the family and to take care of small children.

All these challenges merit the utmost concern prior to the fact that teachers and students are able to realign their commitment back to the restructured and revamped program in the form of online learning. While the problems faced by the students may not be experienced to the similar degree by all higher education populations across the world, generally these trials are collective apprehensions for almost all students as well as teachers.

3.7 CONTRIBUTIONS OF THE STUDY

First and foremost, there is a lack of study into how educational institutions try to deal with the participation of students during any pandemic and the worldwide closing of educational institutions. This research would help to uncover key themes and add to native literature in this regard, which may be employed by concerned administrators to enhance their educational initiatives. By disclosing the benefits of technology integration in teaching/learning, this review also makes a significant contribution through its analysis. Notably, this analysis will be of considerable benefit to the faculty since they are in regular communication with the students and are able to appreciate their actions effectively and discuss problems of online management cordially. In a similar way, valuable details on the advantages of ICT blended learning will also be provided to education policymakers, encouraging them to be included as pedagogical improvements in the education sector. As a way of improving their teaching skills, higher education faculty should understand the value of conducting their studies in online modes using technology. By providing those with valuable insights into ICT integrated instruction, this exploration bounces to the advantage of colleges and universities, helping them to develop their curriculums to effectively train faculty to cope with the various necessities of this pandemic. In particular, they will have to redesign their curriculum so that at the primary and secondary levels, ICT information is included in their text. This adaptation would help train the students at HEIs for ICT integrated pedagogy. Subsequently, in the light of the nation-wide lockdown owing to this pandemic, colleges and universities are able to establish an appealing and fun learning environment for all students. The global importance and policy consequences of this paper lie in its potential to include a realistic and contextual policy guide for HEIs in the development of an Education Continuity Strategy that describes the protocols and guidelines that schools must follow in the face of a pandemic.

3.8 CONCLUSION

Extraordinary times demand exceptional leadership and behavior. In response to Covid-19, planning to transfer schooling beyond conventional physical classrooms requires thinking, preparations, deliberate decision-making and a major transition of learning. As the literature requires an extraordinary degree of student success and participation in an ICT engrossed learning setting, we need to be hopeful. Clearly, technology has had a great impact on the lives of young people in changing the educational environment worldwide. The educational goals and desires of students who have become digital addicts can be synergized by this digitalized movement. To extract this pandemic has given us the ability to accept e-learning, hence educational institutions require to be in line with the swift proliferation of emerging technology, making online, mixed, and distance learning a requirement not only in India but worldwide.

In addition to it, the pandemic and social distance requirement of Covid-19 has posed unnecessary obstacles for all participants to go virtual and perform in a situation of time and resource restrictions. It must be recognized that it is not just a technological problem to accept the online learning environment but also an educational and pedagogical challenge. As such, comprehensive planning with respect to instructional materials and instruction and evaluation information is essential in online education. Technology is the medium of distribution which demands that educational, information and technology departments work closely together. A pedagogical transformation involves closing of universities and colleges and making accelerated deployment of both academic personnel and resources.

Looking at how nations have reacted to the global pandemic so far reveals how each nation has attempted to cope with what is an emergency crisis, which is effectively unparalleled in the history of universal free public education. In these uniquely difficult times, educational systems have been able to offer a sort of consistency in schooling and higher education, primarily mediated by technology and the internet, demonstrating the professionalism and dedication of teachers, professors and educational leaders, and the care and dedication of parents especially those engaged in enforced education at home.

The condition of Covid-19 has pushed students and faculty through online classes into a new realm of remote or distance learning. In order to achieve shared and substantive collective learning, these accelerated changes from conventional face-to-face classes to interactive classrooms have brought about a number of considerations that faculty and students must discuss and come to a consensus. The need for cooperation between students and professors is important, even more so with adult learners, as versatility must be possible to effectively involve all parties across the online forum used (Joiner, 2004).

As a consequence of spread of Covid-19, the severe initiatives, including the closing of schools and colleges, social distancing and the change from face-to-face teaching and studying to e-learning, have impacted millions of students around the world. These reforms not only caused students and educators a great degree of inconvenience but also provided a snapshot of how education could evolve quickly and prompted educational leaders and administrators to look for new models of imagination and progress of education. Although most students worldwide have been able to find an alternative to face-to-face instruction, as of April 2020, only about 60% (Statista, 2020) were active internet users, which may further expand the difference in the standard of education worldwide for students residing in less developed countries. HEIs around the globe have raced to successfully introduce their online learning in response to the Covid-19 crisis. The core interventions that helped universities and colleges sustain their learning cycle while safeguarding the learning trajectory of students were primarily related to swift management and policy measures and access to capital. However, the lack of the past research on similar pandemic challenges, the confusion as to when new therapies and vaccines will be available, has only added to the difficulties in forecasting likely future remedies. A more collaborative, yet strategic, education and research policy is required in the Indian context. The first step in this direction will be for universities, in addition to raising the participation of industry and the private sector, to further deepen their cooperation in teaching, science, collaborative funding and community service in order to ensure successful reverse linkages between universities and the business world. To build new networks between HEIs in the world, Quality Assurance frameworks need to be developed to their maximum strength. For the sake of guiding students and society at large, it is also necessary for HEIs to pay more attention to the Indian sense of job market analysis and developments therein. In conclusion, the development of well-structured, evidence-based online learning environments is a significant next step: not only to involve students in remote education optimally but also to plan for a more or less normalized, blended or hybrid learning situation in schools and universities.

LIST OF ABBREVIATIONS

ICT	Information and Communication Technology
HEIs	Higher Education Institutions
UNESCO	United Nations Educational, Scientific and Cultural Organization
SARS	Severe Acute Respiratory Syndrome
OER	Open Education Resources
MOOCs	Massive Open Online Courses

MHRD	Ministry of Human Resource Development
DTH	Direct-To-Home
UGC	University Grants Commission
PGs	Paying Guests

REFERENCES

Al-Emran, M., Elsherif, H. M., & Shaalan, K. (2016). Investigating attitudes towards the use of mobile learning in higher education. *Computers in Human Behavior*, *56*, 93–102. https://doi.org/10.1016/j.chb.2015.11.033

Ali, W. (2019). The efficacy of evolving technology in conceptualizing pedagogy and practice in higher education. *Journal of Higher Education Studies*, *9*(2), 81–95. https://doi.org/10.5539/hes.v9n2p81

Ali, W. (2020). Online and Remote Learning in Higher Education Institutes: A Necessity in Light of COVID-19 Pandemic. *Higher Education*, *10*(3), 16–25. https://doi.org/10.5539/hes.v10n3p16

Aubusson, P., Schuck, S., & Burden, K. J. (2009). Mobile learning for teacher professional learning: benefits, obstacles and issues. *Research in Learning Technology*, *17*(3), 233–247. https://doi.org/10.3402/rlt.v17i3.10879

Aydin, S. (2012). A review of research on Facebook as an educational environment. *Educational Technology, Research and Development*, *60*(6), 1093–1106. http://dx.doi.org/10.1007/s11423-012-9260-7

Basilaia, G., & Kvavadze, D. (2020). Transition to Online Education in Schools during a SARS-CoV-2 Coronavirus (COVID-19) Pandemic in Georgia. *Pedagogical Research*, *5*(4), em0060. https://doi.org/10.29333/pr/7937

Becker, H. J. (2000). Who's wired and who's not: Children's access to and use of computer technology. *The Future of Children*, *10*(2), 44–75. https://doi.org/10.2307/1602689

Brahimi, T., Khalifa, S., Benaouda, B. (2019). Integrating Makerspaces in Higher Education: Constructionism Approach to Learning. *Research & Innovation Forum 2019 Springer Proceedings in Complexity*, 65–73. doi:10.1007/978-3-030-30809-4_7.

Brahimi, T., & Sarirete, A. (2015). Learning outside the classroom through MOOCs. *Computers in Human Behavior*, *51*, 604–609. doi:10.1016/j.chb.2015.03.013.

Buabeng-Andoh, C. (2015). ICT usage in Ghanaian secondary schools: teachers' perspectives. *The International Journal of Information and Learning Technology*, *32*(5), 300–312. https://doi.org/10.1108/IJILT-09-2015-0022

Buabeng-Andoh, C., & Totimeh, F. (2012). Teachers' innovative use of computer technologies in classroom: A case of selected Ghanaian schools. *International Journal of Education and Development using Information and Communication Technology*, *8*(3), 22–34.

Buabeng-Andoh, C., & Yidana, I. (2014). An investigation of secondary school students' attitudes toward pedagogical use of ICT in learning in Ghana. *Interactive Technology and Smart Education*, *11*(4), 302–314. https://doi.org/10.1108/ITSE-10-2013-0024

Coll, C., Mauri, T., & Onrubia, J. (2009). Towards modeling of the teaching –learning mediated by ICT. *Educational Technology, Teacher education in the Internet age*, 145–161.

Corbeil, J. R., & Valdes-Corbeil, M. E. (2007). Are you ready for mobile learning? *Educause*, *30*(2), 51. Retrieved from https://er.educause.edu/articles/2007/4/are-you-ready-for-mobile-learning

CoSN. (2020). *COVID-19 Response: Preparing to Take School Online*. Retrieved from https://www.cosn.org/sites/default/files/COVID-19%20Member%20Exclusive_0.pdf

Crompton, H., Burke, D. J. C., & Education. (2018). The use of mobile learning in higher education: A systematic review. *Computers & Education*, 123, 53–64. https://doi.org/10.1016/j.compedu.2018.04.007

Czerniewicz, L. (2020). What we learnt from "going online" during university shutdowns in South Africa. Retrieved from https://philonedtech.com/what-we-learnt-from-going-online-during-university-shutdowns-in-south-africa/

Durden, W. (2020, April 08). *Turning the Tide on Online Learning*. Retrieved from Inside Higer Ed: https://www.insidehighered.com/views/2020/04/08/online-learning-can-only-be-viable-if-it-offers-certain-connection-points.

Earle, R. S. (2002). The integration of instructional technology into public education: Promises and challenges. *Educational Technology & Society*, *42*(1), 5–13.

Elisha, O.M., Odini, C., & Ojiambo, J. B. (2013). Use of information communication technologies in education and training of undergraduate library and information science students in two selected Kenyan universities. *Library Review*, *62*(8/9), 585–601. http://dx.doi.org/10.1108/LR-08-2012-0083

Ertmer, P. A., & Otternbreit-Leftwich, A. T. (2019). Teacher technology change: How knowledge, confidence, beliefs, and culture intersect. *Journal of Research on Technology in Education*, *42*, 255–284. https://doi.org/10.1080/15391523.2010.10782551

Fu, J. S. (2013). ICT in education: A critical literature review and its implications. *International Journal of Education and Development using Information and Communication Technology*, *9*(1), 112–125.

Fullan, M. (2013). *Stratosphere: Integrating technology, pedagogy and change knowledge*. Pearson Education.

Geng, S., Law, K., & Niu, B. (2019). Investigating self-directed learning and technology readiness in blending learning environment. *International Journal of Educational Technology in Higher Education*, *16*. https://doi.org/10.1186/s41239-019-0147-0

Haddad, W. D. (2003). *Is instructional technology a must for learning?* Retrieved from http://www.techknowlogia.org/TKL_active_pages2/CurrentArticles/main.asp?IssueNumber=19&FileType=HTML&ArticleID=455

Hennessy, S., Deaney, R., & Ruthven, K. (2006). Situated expertise in integrating use of multimedia simulation into secondary science teaching. *International Journal of Science Education*, *28*(7), 701–732. https://doi.org/10.1080/09500690500404656

Huang, H. M., & Liaw, S. S. (2005). Exploring user's attitudes and intentions toward the web as a survey tool. *Computers in Human Behavior*, *21*(5), 729–743. https://doi.org/10.1016/j.chb.2004.02.020

ILO. (2020, April 20). *COVID-19 and the education sector*. Retrieved from https://www.ilo.org/sector/Resources/publications/WCMS_742025/lang--en/index.htm

Jena, P. K. (2020). Challenges and opportunities created by Covid-19 for ODL: A case study of IGNOU. *International Journal for Innovative Research in Multidisciplinary Field*, 6(5): 217–222.

Jesse, G. R. (2015). Smartphone and app usage among college students: Using smartphones effectively for social and educational needs. *Issues in Information Systems*, 17(4), 8–20.

Joiner, R. (2004). Supporting collaboration in virtual learning environments. *Cyber Psychology & Behavior*, 7(2), 197–200.

Jung, I. (2005). ICT-pedagogy integration in teacher training: Application cases worldwide. *Journal of Educational Technology & Society*, *8*(2), 94–101.

Lanzilotti, R., Ardito, C., Costabile, M. F., & De Angeli, A. (2006). eLSE methodology: a systematic approach to the e-learning systems evaluation. *Educational Technology & Society*, *9*(4), 42–53.

Li, S., Yamaguchi, S., & Takada, J. (2018). Understanding factors affecting primary school teachers' use of ICT for student-centered education in Mongolia. *International Journal of Education and Development using Information and Communication Technology*, *14*(1), 103–117.

Michelsen, G., & Wells, P. J. (2017). *A decade of progress on education for sustainable development: Reflections from the UNESCO chairs programme*. UNESCO Publishing.

Miks, J., & McIlwaine, J. (2020). Keeping the world's children learning through COVID-19. Retrieved from UNICEF: https://www.unicef.org/coronavirus/keeping-worlds-children-learning-through-covid-19

Mirzajani, H., Mahmud, R., Ayub, A. F. M., & Wong, S. L. (2016). Teachers' acceptance of ICT and its integration in the classroom. *Quality Assurance in Education, 24*(1), 26–40. https://doi.org/10.1108/QAE-06-2014-0025

Moopen, A. (2020). Mental health in the time of COVID-19 outbreak. Retrieved from Gulf News: https://gulfnews.com/opinion/op-eds/mental-health-in-the-time-of-covid-19-outbreak-1.70680268

Murgatrotd, S. (2020). *COVID-19 and Online Learning*. doi:10.13140/RG.2.2.31132.85120

Naciri, A., Baba, M. A., Achbani, A., & Kharbach, A. (2020). Mobile learning in higher education: Unavoidable alternative during COVID-19. *Aquademia, 4*(1), ep20016. https://doi.org/10.29333/aquademia/8227

Ruzgar, N. S. (2005). A research on the purpose of internet usage and learning via internet. *The Turkish Online Journal of Educational Technology, 4*(4), 27–32.

Sadegül Akbaba, A., Kalayci, E., & Avci, Ü. (2011). Integrating ICT at the faculty level: A case study. *TOJET: The Turkish Online Journal of Educational Technology, 10*(4), 230–240.

Sahu, P. (2020). Closure of universities due to coronavirus disease 2019 (COVID-19): Impact on education and mental health of students and academic staff. *Cureus*, 12(4), 1–6.

Sha, L., Looi, C. K., Chen, W., & Zhang, B. H. (2012). Understanding mobile learning from the perspective of self-regulated learning. *Journal of Computer Assisted Learning, 28*(4), 366–378.

Shachar, M., & Neumann, Y. (2010). Twenty years of research on the academic performance differences between traditional and distance learning: Summative meta-analysis and trend examination. *MERLOT Journal of Online Learning and Teaching*, 6(2), 318–334.

Sipilä, K. (2011). No pain, no gain? Teachers implementing ICT in instruction. *Interactive Technology and Smart Education, 8*(1), 39–51. http://dx.doi.org/10.1108/17415651111125504

Smith, J. A., & Judd, J. (2020). COVID-19: Vulnerability and the power of privilege in a pandemic. *Health Promotion Journal of Australia, 31*(2), 158–160. http://dx.doi.org/10.1002/hpja.333

Solomon, E. (2020). *Khalifa University's Response to Covid-19*. Retrieved from https://www.ku.ac.ae/khalifa-universitys-response-to-covid-19

Sönmez, A., Göçmez, L., Uygun, D., & Ataizi, M. (2018). A review of current studies of mobile learning. *Journal of Educational Technology and Online Learning, 1*(1), 12–27. https://doi.org/10.31681/jetol.378241

Statista. (2020). *Global digital population as of April 2020*. Retrieved from Statista: https://www.statista.com/statistics/617136/digital-population-worldwide.

Steel, C. (2012). *Fitting learning into life: Language students' perspectives on benefits of using mobile apps*. Paper presented at the *Future challenges, sustainable future, Proceedings of ascilite conference Wellington 2012*.

Stern, A.M., Cetron, M.S. & Markel, H. (2009). Closing the schools: lessons from the 1918-19 US Influenza pandemic. *Health Affairs, 28* (S1), 1066–1078.

Thomas, O. O., & Israel, O. O. (2013). Effectiveness of animation and multimedia teaching on students' performance in science subjects. *British Journal of Education, Society & Behavioural Science, 4*(2), 201–210. https://doi.org/10.9734/BJESBS/2014/3340

Toquero, C. M. (2020). Challenges and Opportunities for Higher Education amid the COVID-19 Pandemic: The Philippine Context. *Pedagogical Research, 5*(4), em0063. https://doi.org/10.29333/pr/7947

UNESCO. (2015). *Qingdao Declaration* (Seize digital opportunities, lead education transformation) (pp. 1–54). Qindgao, China. Retrieved from http://unesdoc.unesco.org/images/0023/002333/233352m.pdf

UNESCO. (2020). *COVID-19 Educational Disruption and Response*. Retrieved from https://en.unesco.org/covid19/educationresponse/

UNESCO. (2020b). *Distance learning solutions*. Retrieved from https://en.unesco.org/covid19/educationresponse/solutions

Visvizi, A., Lytras, M.D., & Sarirete, A. (Eds.) (2019). *Management and administration of higher education institutions at times of change*. Emerald Publishing, ISBN: 9781789736281, https://books.emeraldinsight.com/page/detail/Management-and-Administration-of-Higher-Education-Institutions-in-Times-of-Change/?K=9781789736281

White, C. P., Ramirez, R., Smith, J. G., & Plonowski, L. (2010). Simultaneous delivery of a face-to-face course to on-campus and remote off-campus students. *TechTrends*, *54*(4), 34–40.

Willms, J. D., & Corbett, B. A. (2003). Tech and teens: Access and use tech and teens: Access and use. *Canadian Social Trends*, 69, 15–20.

World Bank. (2020). *Remote Learning and COVID-19*. Retrieved from https://documents1.worldbank.org/curated/en/266811584657843186/pdf/Rapid-Response-Briefing-Note-Remote-Learning-and-COVID-19-Outbreak.pdf

4 Investigating Academic Transition during the COVID-19 Pandemic
The Case of an Indian Private University

Upasana G Singh
University of KwaZulu-Natal, South Africa

Dilip Kumar Sharma
GLA University, India

CONTENTS

DOI: 10.1201/9781003102298-4

4.1 INTRODUCTION

The higher education (HE) scenario has changed drastically worldwide due to the COVID-19 pandemic. HE transformed overnight to adopt new trends like increasing online education, increasing competition between universities, and embracing digital technologies. The main aim of digital learning is to increase students' productivity to utilize their time more efficiently and effectively without any loss of their study (Mugo, Odera & Wachira, 2020).

In this pandemic situation, online learning was considered the best way to support the continuity of academic programs in Higher Education Institutions (HEIs) across the world (Cape Argus, 2020; USAF, 2020; Yamin, 2020). During the crisis of COVID-19 HEIs worldwide were forced to adopt new institutional polices and had to rethink their institutional planning for staff and students (USAF 2020).

This research considers the attitudes of academics at a HEI in India who had to transition from classroom learning to online/virtual learning platform during the COVID-19 epidemic. This study's main focus was to understand the experiences faced by the academics at this selected HEI in this pandemic situation. In the state of COVID-19 the staff were forced to adapt the new mode of education, i.e. the digital one, where they were responsible for providing continued education. In this study we focused on the following points:

- Understanding the Online Learning Tools used to support online teaching and assessments
- Determining the perception of challenges faced by students in moving online
- Understanding communication methods adopted during WFH
- Recognizing the general impact of COVID-19 on academics

4.1.1 The New Trends of Education Systems

4.1.1.1 Online Learning

The new trend in education is the adoption of digital learning by using various platforms, such as Byju's, Udemy, Goodreads, edX, and Zoom. In the future, it is possible that if there is a positive impact of online learning, this may lead to eliminating the role of class-based learning.

There are various constraints to online learning (Mugo, Odera & Wachira, 2020) which include internet access, the cost of data, and access to electricity, and the physical and social setup, which is often not conducive to learning (Mdepa, 2020; Mthethwa, 2020; Mzileni, 2020). For many students, the HEI physical facility is a haven, where their various needs (accommodation and access to facilities, including the library and computer LANs) are met.

Despite the challenges associated with online learning, the COVID 19 pandemic also provided various opportunities for both students and staff in the form of webinars, online workshops, online expert talk, online tutoring, e-assessment, and many more that will help transform the education system in the new phase (Alruwais, N., Wills, G., &; Wald, M. (2018). This experience validated the work of Appavoo, Sukon, Gokhool, and Gooria (2018) who presented the various advancements in technology that enhance the online learning system in HEIs.

4.1.1.2 Comprehensive and Digital Online Assessments

Before COVID-19, the examination pattern/system was different as students wrote the theoretical answers in the answer sheet, which were then evaluated by the evaluator. Now, the online assessment is quite tricky with the use of technology. Artificial Intelligence (AI) plays an essential role in the new learning system. The online examination will be conducted with the help of technology. This system can provide a beneficial insight by which the student performance, individual scrutiny, and group scrutiny for each topic can be analyzed. The new AI tools and techniques are used, which reduces the classroom-based system problems like human dependency, biasness, error, social behavior, etc. There will be more emphasis in the coming years on student assessment for the practical knowledge of the particular subject. This would reduce the proportion of theory tests. Online test program such as Eklavvya.com also offers subjective examination assessment.

4.1.1.3 Personalized Education

Personalized learning overcomes the limitation of traditional knowledge. Nowadays traditional classroom facility is becoming outdated. Students like to learn using technology mode. Personalized education technique also depends on the liking of individual students on a particular topic or subject. One of the patterns that will rule over the coming year is personalized learning using technology. It overcomes conventional learning characteristics such as reliant on the classroom, fixed location, etc.

4.1.1.4 Revolution in Exam Management

As we know earlier, the examination pattern was different, but the trend which is going on is eliminating the traditional method and adopting the AI-based examination method, which will help in conducting exams on time with proper security. AI-based proctoring or auto remotes proctoring will allow colleges/universities to conduct exams without any hurdle, i.e. requirement of infrastructure or logistics. The new feature of remote proctoring technology helps the students appear in the exam at any location, just with the internet's help. AI-based remote proctoring technology will be on demand in the coming days.

4.1.2 Corona Virus Impact on Learners

The COVID situation adversely impacted on the mindset of students (Gutterer, 2020). Although earlier the students usually were likely to go abroad for their HE, things have changed. According to the study portal survey on students, they came to find out that 7% of respondents expected to enroll within the next six months, that is, if not sooner, to add meaning to the study in the upcoming winter semester. Other groups' timeframe includes: 27% plan to register in year 2020, 12% plan to register within 12 to 18 months; 11% plan to begin studying later. Students used internet to fetch information and compared the existing options. The significant change came for them after they decided to study abroad even before they submitted applications.

According to the Association of Indian Universities (AIU) (Pushkarev, 1975): The epidemic of COVID 19 is damaging the economy and health and destruct the education system (Pushkarev, 1975). Thus, many universities, institutions, schools, colleges closed, and the unemployment ratio is continuously rising. The closure of educational institutions for a long time is having a severe impact on education. Due to this, there is a loss of academic knowledge examination. To cope up with this, government has initiated the method of online classes. They have also started 22 new channels for the students so that their regular study is not disturbed and they can do their study at their home without going anywhere.

Therefore, the AIU has taken the lead and made an association where the members of various institutions, universities from India, share the e-content, strategies they have adopted to prevent educational loss. The association also takes the suggestions and inputs to choose and share the best possible practices to minimize the effect of this pandemic.

4.1.3 COVID-19 and Online Teaching in HE Instructional Strategies for Future

4.1.3.1 Making Emergency Plans for Unexpected Hurdles in Education

Now, the concept of online education evolving day by day can create a burden on the database server. To address this overload situation, the teachers must have an alternative in the form of Plan B or Plan C to handle any problem that occurred during the lecture.

4.1.3.2 Divide the Study Material to Smaller Units

Sometimes, online learning can create a disturbance. Students can suffer from weak consistency in online learning (Li et al., 2013). To ensure that students can learn new concepts and maintain consistency, they must make an interactive video lecture session. There is a need to divide the lectures module-wise so that it can help develop students' interest. Lecture timing should not be more than 20–25 minutes.

4.1.3.3 Face-to-Face Interaction in Online Classes

In conventional teaching mode, study material, body language, facial expressions, and voice of the teacher play an essential role while delivering the lecture. But in an online class, the only voice of the teacher and study material provides a concept to students. So it is required to maintain the pace of speech to allow students to grasp the main ideas.

4.1.3.4 Ensure Coordination with Faculties and Gain Full Online Support From Them

Faculty must provide help to students during the epidemic situation. There is a need to use online educational platforms in a better way. For new faculty members, the technical criteria of online education are much higher than conventional in-class instruction. Given that most of our university's faculty is insufficiently qualified or funded to operate online educational platforms, teaching assistants' support is significant. Before the lesson, the faculty will thoroughly consult with the teaching assistants to ensure that they understand each student's goals, information structure, and teaching activities.

4.1.3.5 Generate Effective Learning Ability Outside the Online Class

Faculty has less influence over online teaching than conventional in-person lectures. There is a need to incorporate various approaches to provide teaching to students.

Thus for academics to successfully transition to the online environment, a new academic skillset has to be developed through the facilitation of training by HEIs (Singh, 2020b).

4.2 RESEARCH PROBLEM

The COVID-19 pandemic forged a sudden and dynamic shift for academic staff who were 'forced' to adopt online teaching and assessment tools to continue their students' academic activity. Thus, this chapter investigates academics' experiences during this transition to online teaching and measures the effects of the unprecedented disruption at this private HEI institution. To understand the impact of this transition, the aim of this study was threefold; to

- Understand academics' preferences of technology to support online teaching and assessment at this private HEI
- Determine academics' experiences with the shift to a forced 'work-from-home' (WFH) situation at this private HEI
- Investigate the general impact of the pandemic on academics at this private HEI

4.3 LITERATURE REVIEW

4.3.1 Impact of COVID-19 in India

In some states like UP, Delhi, etc., it was announced to shut down schools and colleges from 13 March 2020 (Mansoor, 2020; The Economist, 2020) (Worldometers, 2020; Huelsman, 2020). The Maharashtra government announced closing the schools on 16 March, and they also postponed all university examinations (Mansoor, 2020). As per the University Grants Commission (UGC) guidelines, Pondicherry University announced closing all the academic activities on 17 March (TNN, 2020; ET Government, 2020). The University of Hyderabad communicated this on March 20 and it would suspend all the academic activities as well as the governments also close the hostels. The same was followed by the many other institutions like SRM Institute, VIT-Vellore they declared the summer vacation for students (Sujatha & Chatterjee, 2020). Till March, no university announced the use of online classes because of COVID-19, and they postpone all their academic activities until the end of March.

Coronaviruses have a high impact on HE. Now the education scenario has been totally changed. The institutions have begun online classes, online assessments for students' welfare so that their studies, education, and a semester cannot be disturbed (Gupta & Goplani, 2020). This is a crucial matter for all, so many colleges and institutions started to take a preventive measure to keep their staff, faculties, and students healthy and make a plan to keep education going from any location by adopting the government measures. As per the situation, Universities and institutions will continue to carry the education through online mode and have to deal with all these challenges in the near term (Huelsman, 2020).

Access to technology and internet connectivity are major challenges in developing countries as highlighted by Allo (2020). Houlden and Veletsianos (2020) indicate that HEIs are not prepared for the digital era of learning due to internet access limitations in many countries. Queiros and de Villiers (2016: 179) highlight the importance of social presence, noting 'interaction with (the) facilitator' and 'timely feedback from (the) facilitator' as important. The presence of the lecturer is critical in online learning (Butcher, 2014). Students show less interest in online learning in comparison to offline learning (Blackmom & Major 2012).

The financial impact of the epidemic on the Universities (Wenham et al., 2020)—Universities that fail to adapt their teaching process to online education could be at the risk of permanent closure. But some institutions could get benefit from this situation. Several universities in India have shut campuses and switched to online mode to provide online classes. Already it creates a negative impact on faculties and staff. Some institutions have cut their staff members' salaries by more than 40%, and some are planning for cost-cutting by a ramp down the resources. School closures in China, Italy, South Korea, and elsewhere to control the transmission of COVID-19 could have a significant impact on women who provide informal care in communities, restricting their job opportunities in the economy.

An effective online learning/ education system must require investment in an ecosystem for students support and not merely just online content delivery. Blewett and Hugo (2016) believe that face-to-face approaches are not necessarily effective online. What is required is a careful consideration of technology's affordances, which need to be viewed through an appropriate pedagogic lens to ensure effective teaching (Blewett & Hugo, 2016).

4.3.2 Challenges Faced by Universities Across the Globe

Shifting of classroom learning to online classes (Lim, 2020), (Sahu, 2020): The online mode's move excited many teachers and students worldwide. The teachers have already started preparing presentations for online delivery of lectures. In many Universities, online teaching existed before COVID-19 pandemic. Many faculty, staffs are trained in using online learning platforms. Nonetheless, for some faculties, this online method is quite difficult for them as they are not techno-savvy (Singh, 2020a).

Shifting of classroom assessment and evaluation toward online assessment and evaluation (Dill et al., 2020): In the era of COVID-19, we found that various and most of the universities have suspended or postponed the final examinations for the semester-end, while with the help of new Learning Management Systems (LMS) the digital classes were continued. It is a challenging task to put online tests into those lessons designed for teaching in the classroom. Both students and teachers are unsure of the process related to the assessment of assignments, projects and other ongoing learning activities. Students without internet connection will be adversely impacted and find difficulties during assessment process, which can affect their results also.

Students in Business Schools in South Africa who pursue part-time postgraduate studies are typically more mature are most often employed and tend to have higher levels of motivation (Wilson, 2018; Russell, 2019; Schwartz, n.d.). Their employers may fund them, and most often, they have their personal system and good speed internet connection. So, in theory, this cohort of students should be in a better position to transition to online learning, thus offering an opportunity to explore other vital considerations that may also impact the shift to online learning. However, Singh, Proches, Leask, and Blewett (2020) identified that for adult learners to experience a successful transition to the online environment, the factors of social interaction, conducive study space, social support, and technological competence are key.

In this pandemic situation, it is quite difficult for many students studying abroad to return their home country (Bothwell, 2020). To ensure food, lodging, and security service for these non-national students has emerged as a great challenge for administrators. Students also need proper protection against any contact with the individual and live in self-isolation until the situation becomes completely normal. Extension of accommodation due to exam delays can create a financial crisis. Again, the disruption resulting from COVID-19 can impact foreign student admissions for the upcoming academic session (Timmis et al., 2016).

According to Jandrić et al., 2018 (Zhu and Liu, 2020), the Coronavirus outbreak has significantly increased the growth of all nation's HE online. S helps in making transition from offline mode to online mode. National post-pandemic teacher education may consist of man-to-man staff education, integrated faculty training, and online faculty education (Zhu and Liu, 2020) (Xudong, 2020).

4.3.3 Post COVID-19 Scenario

The way we teach and assess needs to be overhauled post the COVID pandemic (ET Government, 2020).

In India, 40 million students enrolled in HE, spread across the country in over 45,000 colleges and nearly 1000 universities. The gross HE enrollment ratio is still only 26, showing, as it is, the amount of space to be generated for the remaining 74 out of every 100 qualified. However, scaling those numbers presents tremendous challenges. Post Corona, what? For everyone, it will be a meaningful and consequential time. It is a given that the academic work remains unfinished, that is, it could not be completed in HEI universities by the end of February 2021 unless innovative ideas emerge. Collaborations on Online learning sites with different reputable online educators must be vigorously sought in order to complete the remaining academics.

Universities just need to be initiators and enablers. Nothing more. They allow technical platforms to offer online training-evaluation processes based on preference and online training. They have to allow boot camps. Currently, the programs are cast in stone and can't innovate. The learning mode should never be a constraint. The universities now need to promote online learning like anything else that shifts. A cognizant step toward a blended learning model needs to be encouraged.

Dr. S Mantha, former Chairman of AICTE and the new Chancellor of KL University, India, talks on various topics with Brainfeed Higher Education Plus. One work in turbulent times...but 'you still don't hear anything when you shut one eye.' It's a wakeup call!! This virus assumes colossal proportions worldwide, with more than 7.2 lakh active cases reported with the number increasing by the day, with nearly a whopping 31,000 dead with recovery in sight nowhere. It has disrupted every viable business that affects the economy.

Schools need to switch to online instruction and teaching. It's simple; the teachers now record their lessons on video at their own homes and send them to WhatsApp students. That will also prepare a digital content library for the schools. Students in the class can visit the content often, rather than for a limited period. This must include all the teachers and support staff as well as the students who must be promptly asked to stay home and learn. Like organizations and businesses, allowing their employees to WFH, our colleges must also enable learning and assessment from home in these extremely distressing times.

Many Internet Providers made their content freely accessible to anyone who wants it, at least until the end of June. In the professional education area, the content of reputable players like Udemy, Coursera, GradeUp can be used by students. Edutech major Byju's said it would give school students free access to its entire app until April. Academic textbooks are also available in the online library Cambridge Core, Cengage, and Wiley. Any institution/college/university shall use the opportunity to reach out to the students to complete the remaining work. It will improve the education process and lower educational costs. The training was not required when using a mobile phone. It's not a battle to switch to Online Learning. In reality, it's enjoyable. Each such effort provides several online resources for fast, free navigation. Therefore, computer literacy and technological issues are non-issues.

Considering today's scenario, it is neither feasible nor convenient to create ever more brick and mortar structures in the future. Peoples will adopt mobile-based learning in the upcoming days. Online learning is independent of time and place. The pattern of exams will, of course, change completely. More weightage should be given to practical knowledge instead of a theoretical one. This pandemic has turned the old chalk-talk teaching model to a technology-based model. Video conferencing app like zoom helps the students to grasp the concept by the teacher.

4.4 RESEARCH METHODOLOGY

This study is part of a broader study on the consequences of the COVID 19 disease on HEIs' academics. Ethics approval was received from the University of KwaZulu-Natal, South Africa, as per Protocol Reference Number HSSREC/00001284/2020.

The focus of this study was on HE academics at the selected private HEI in India. An online survey was developed and distributed through direct email invitation to academic staff at the selected institution to gather primary data. Quantitative data was collected to analyze usage and adoption of technology for teaching, and assessment, both prior to and during the pandemic. Data collection was open for a period of 3 weeks. The online survey took approximately 15 minutes to complete and was anonymous. The final sample achieved was 222 responses. Data were analyzed using statistical analysis using SPSS. Tests used in the analysis comprises of Descriptive statistics through means and standard deviations methods, where applicable. Frequencies are shown in tables or graphs; Chi-square goodness-of-fit-test—a univariate test, used on a thorough variable to test whether any of the response options are selected remarkably more/less often than the others; Chi-square test of independence—to see whether a notable relationship exists between two variables represented in a cross-tabulation. When conditions are not up to the criteria, Fisher's exact test is used; binomial test—tests whether a fundamental proportion of cons select one of the possible two responses; One sample t-test—identifies whether a mean score is abruptly different from a scalar value; Independent samples t-test—a test that compares two independent groups of cases; and Friedman's test—to test the repetitive contents.

4.5 ANALYSIS OF RESULTS

A total of 222 valid responses were collected in this study. Table 4.1 summarizes the demographic distribution of the respondents. Most of the participants (55.4%) were from the ages of 31 to 40. Males (82.0%) dominated the study, with a large

TABLE 4.1
Demographics Summary

Variable/categories	Frequency (%)	Variable/categories	Frequency (%)
Age		**Experience**	
20–30	60 (27.0%)	1–5	71 (32.0%)
31–40	123 (55.4%)	6–10	68 (30.6%)
41–50	33 (14.9%)	11–15	56 (25.2%)
51–60	3 (1.4%)	16+	27 (12.2%)
>60	3 (1.4%)		
Gender		**Academic Role**	
Female	40 (18.0%)	Tutor	2 (0.9%)
Male	182 (82.0%)	Lecturer	64 (28.8%)
		Senior lecturer	37 (16.7%)
Qualification		Associate professor	19 (8.6%)
Undergraduate degree/diploma	24 (10.8%)	Full professor	9 (4.1%)
Postgraduate degree	198 (89.2%)	Assistant professor	84 (37.8%)
Tenure		Other	7 (3.2%)
Permanent	182 (82.0%)		
Contract	40 (18.0%)		

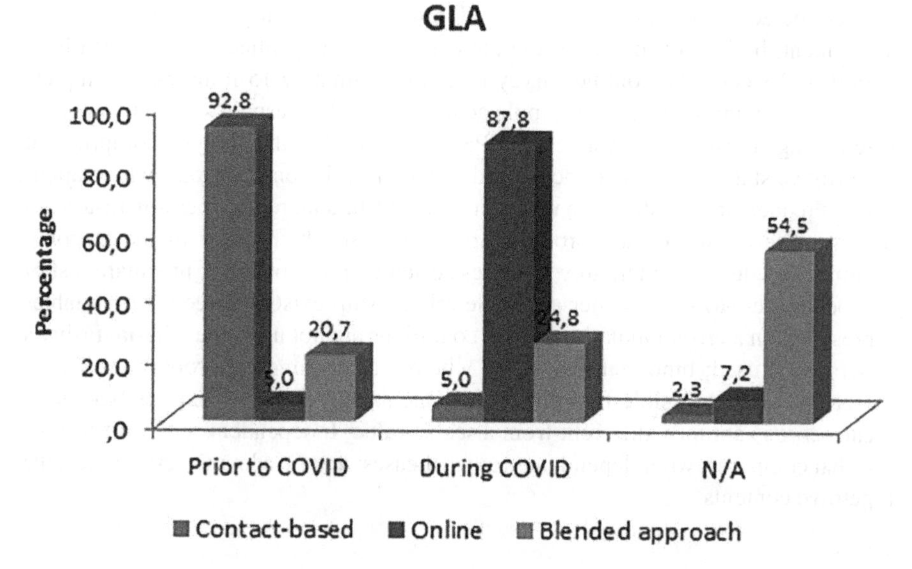

FIGURE 4.1 Primary Teaching Method.

majority of respondents (89.2%) holding a postgraduate degree. Cumulatively, the Professor categories (50.5%) held the highest participation rate, with only 37.4% of participants having experience in academia for more than ten years. The majority (82.0%) had permanent tenure.

To determine the impact that the COVID-19 pandemic had on the primary teaching method, participants were asked to indicate their adoption of the three dominant types, which is contact-based (face-to-face), online learning (no face-to-face), and blended learning. As depicted in Figure 4.1, there was a clear shift from contact-based (92.8%) and blended learning (20.7%) prior to the COVID-19 pandemic to E- learning (87.8%) during the pandemic.

The study then shifted focus to investigate these academics' proficiency in adopting technology to support online teaching and online assessment.

Respondents rated their proficiency in adopting both online teaching and online assessment methods using the scale from 1 = poor to 5 = excellent. A one-sample t-test was applied to test if the average proficiency rating was significantly above or below an average rating of '3'. Results showed that proficiency ratings were significantly above average for both teaching methods (mean rating = 4.26, $p<.0005$) and assessment methods (mean rating = 3.95, $p<.0005$). This indicates that the respondents rated themselves as better than average, tending towards excellence in their proficiency in adopting technology for both teaching and assessment.

Respondents were asked to identify the tools they adopted to support online teaching, prior to and during the pandemic. The results are summarized in Figure 4.2.

As illustrated in Figure 4.2, during the pandemic, there was a clear increase in the adoption of LMS tools like BlackBoard (72.5%), virtual platforms like Zoom

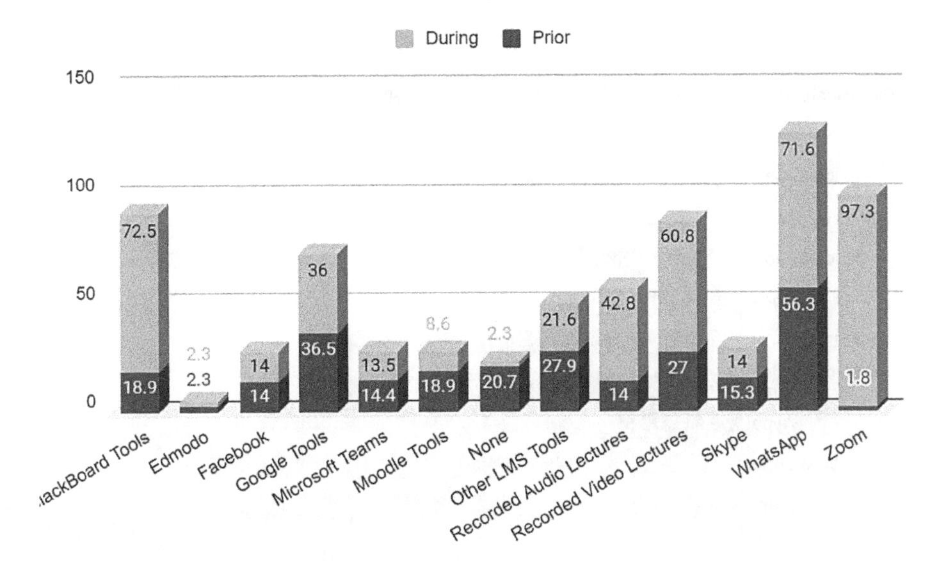

FIGURE 4.2 Teaching Tools Adoption Summary.

TABLE 4.2
Significant Teaching Tools

Teaching Tool	%	p-value
Zoom	97	<.0005
Recorded video lectures	61	.002
WhatsApp	72	<.0005
BlackBoard tools	73	<.0005

(97.3%), and social tools like WhatsApp (71.6%) to support online teaching. There was also a marked increase in the adoption of Recorded Audio Lectures (42.8%) and Recorded Video Lectures (60.8%) for online teaching at this Institute. The number of academics who do not adopt online teaching tools dropped to just 2.3% during the pandemic.

A binomial test was conducted to identify if a significant proportion of the sample responded Yes or No to the usage of each of these tools prior to or during the pandemic. Results showed, as presented in Table 4.2, that Zoom (97%, $p<.0005$); Recorded video lectures (61%, $p=.002$); WhatsApp (72%, $p<.0005$), and BlackBoard (73%, $p<.0005$) were adopted by a significant number of academics during the pandemic.

Similarly, respondents were asked to identify the tools they adopted to support online assessment prior to and during the pandemic. The results are summarized in Figure 4.3.

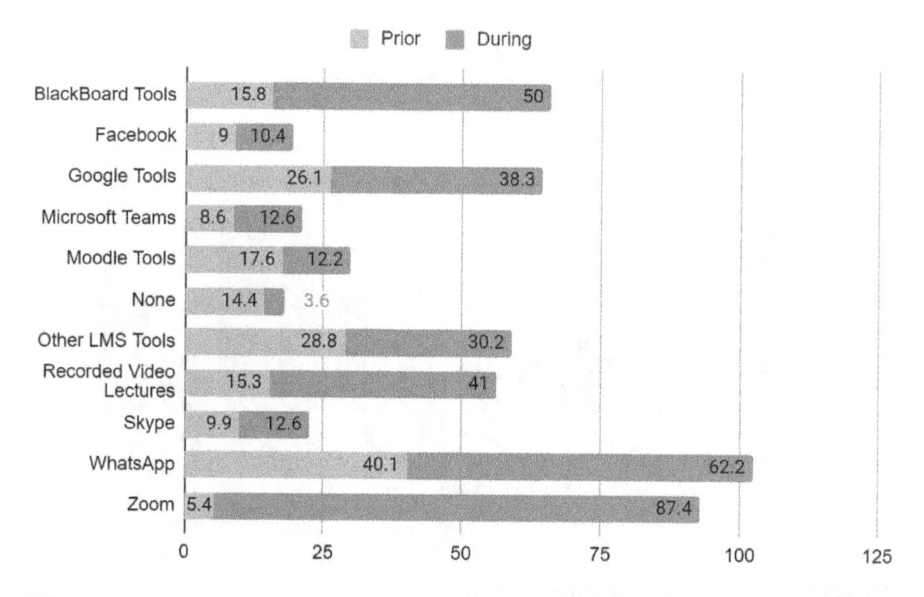

FIGURE 4.3 Assessment Tools Adoption Summary.

As illustrated in Figure 4.3, there was a similar pattern observed in the adoption of online assessment tools during the pandemic, as with online teaching tools. There was a clear increase in the adoption of LMS assessment tools like BlackBoard (50%), virtual platforms like Zoom (87.4%), and social tools like WhatsApp (62.2%) to support online assessment. There was also a marked increase in the adoption of Google Tools (38.3%) for assessment. The number of academics who do not adopt online teaching tools dropped to just 3.6% during the pandemic.

The binomial test results showed such a significant adoption of any of the listed assessment tools by academics at this Institute during the pandemic.

When asked to rate their proficiency in communicating with their students using an online learning platform during the pandemic on a scale from 1 = very difficult to 5 = very easy, no significant results were noted.

From Figure 4.4, it is noted that 'Access to connectivity' (91.9%) was significantly ($p<.0005$) highlighted as the biggest obstacle for students. Thereafter, the major challenges academics at this Institute perceived their students were facing were 'Lack of interaction' (29.7%), and 'Access to devices' (24.3%).

At this institution, academics noted that the most effective methods to support their students during the pandemic are 'Online lectures' (95.9%), 'Online assessment' (82.9%), and 'Phone' (73.9%), as illustrated in Figure 4.5.

These were all indicated by a significant proportion of the respondents as they supported their students ($p<.0005$ in each case), as summarized in Table 4.3.

When rating the responsiveness of their students to the support, on a scale from 1 = unresponsive to 5 = very responsive, the mean rating was 3.97, which is

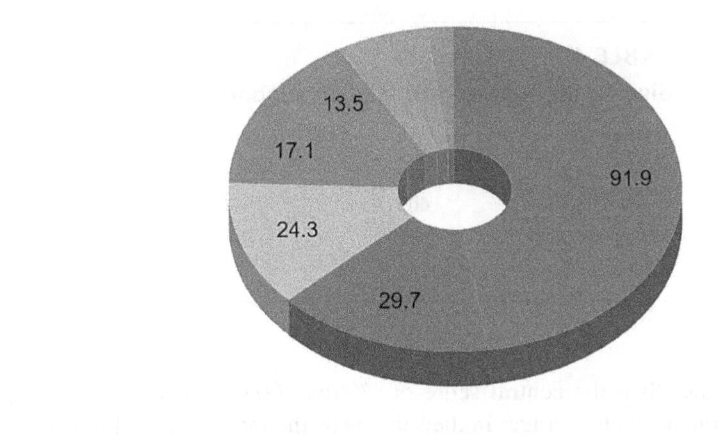

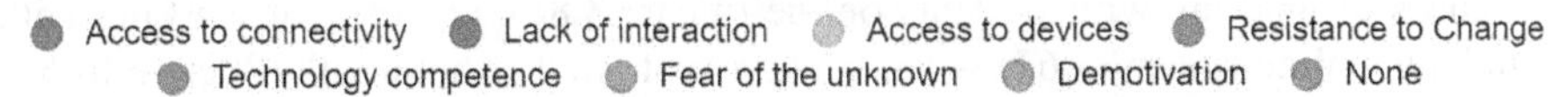

FIGURE 4.4 Perception of challenges faced by students in moving online.

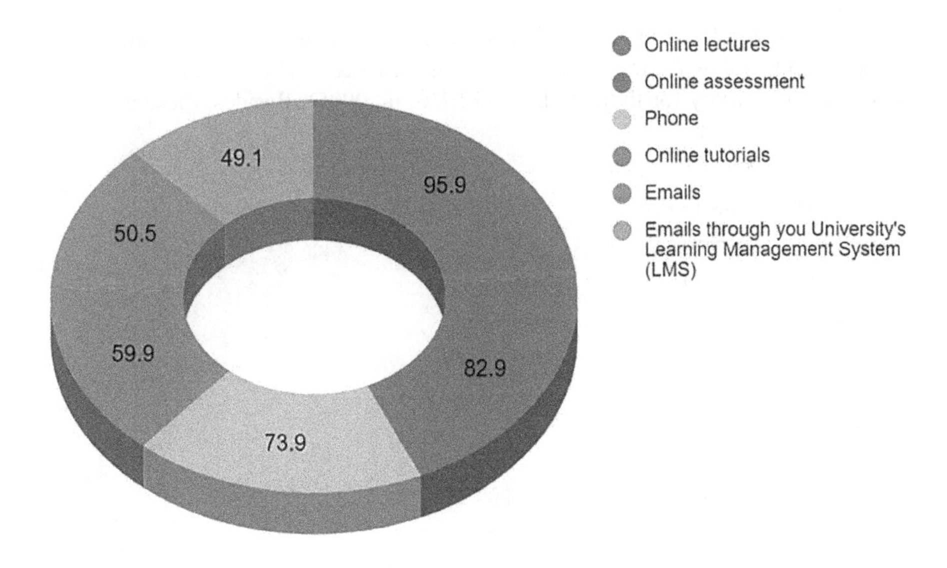

FIGURE 4.5 Methods to support students during the pandemic.

TABLE 4.3
Significant methods for student support

Method	%	p-value
Online lectures	96	<.0005
Online tutorials	60	.004
Online assessment	83	<.0005
Phone	74	<.0005

significantly higher than the central score of '3' (p<.0005), indicating that they were more responsive than average. In the same way, the mean rating of the effectiveness of working with students online (mean=3.88), and ease of working with students online (mean=3.67) when using a scale of 1 = not at all effective to 5 = very effective was also significantly higher than the average score of '3, indicating better than average effectiveness and ease of working (p<.0005). There was a further significant indication that they would continue with online learning post the pandemic (mean=3.13, p<.0005).

Figure 4.6 presents the impact that the shift to work from home had on the lives of these academics.

The most significant challenge that academics at this Institute faced with teaching their students online during the pandemic, in the forced WFH scenario, was that of internet connectivity for their students (72.1%, p<.0005). Besides, these academics

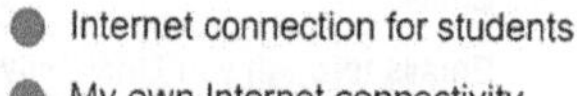

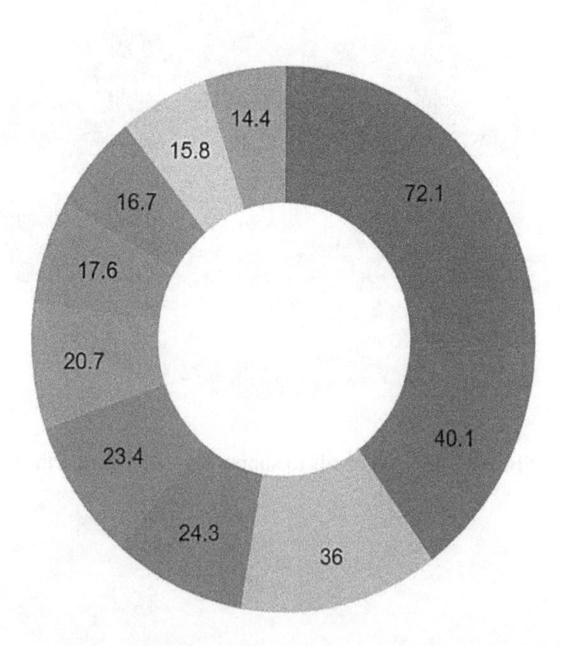

FIGURE 4.6 Challenges with the forced WFH scenario.

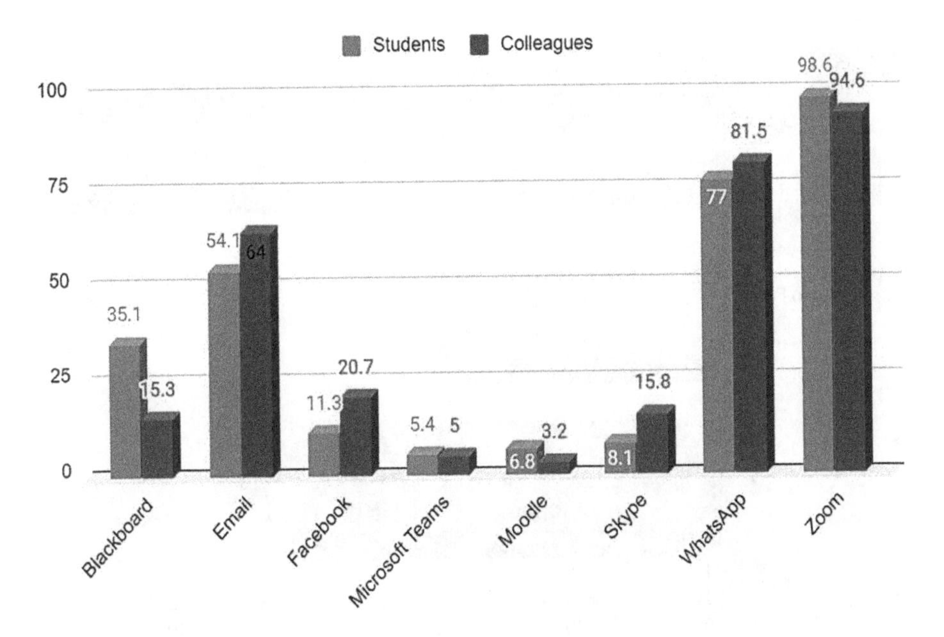

FIGURE 4.7 Communication methods adopted during WFH.

TABLE 4.4
Significant communication tools adopted during WFH

Tool	Students		Colleagues	
Zoom	99	<.0005	95	<.0005
Email	---	---	64	<.0005
WhatsApp	77	<.0005	82	<.0005

(401%) also battled with their own internet connectivity (40.1%) and General anxiety about COVID-19 (36%).

The most popular tools adopted for communicating with students and colleagues in the WFH situation were Zoom, WhatsApp, and Email, as illustrated in Figure 4.7. Academics at this Institution adopted Zoom and WhatsApp significantly more than other communication tools for interaction with their students and colleagues, while Email was also included in this list for communication with colleagues, as summarized in Table 4.4.

Most academics felt that this institution would adopt more online learning (47.1%) than blended learning (31.5%), and even less face-to-face learning (20.8%, post the COVID-19 pandemic, as illustrated in Figure 4.8.

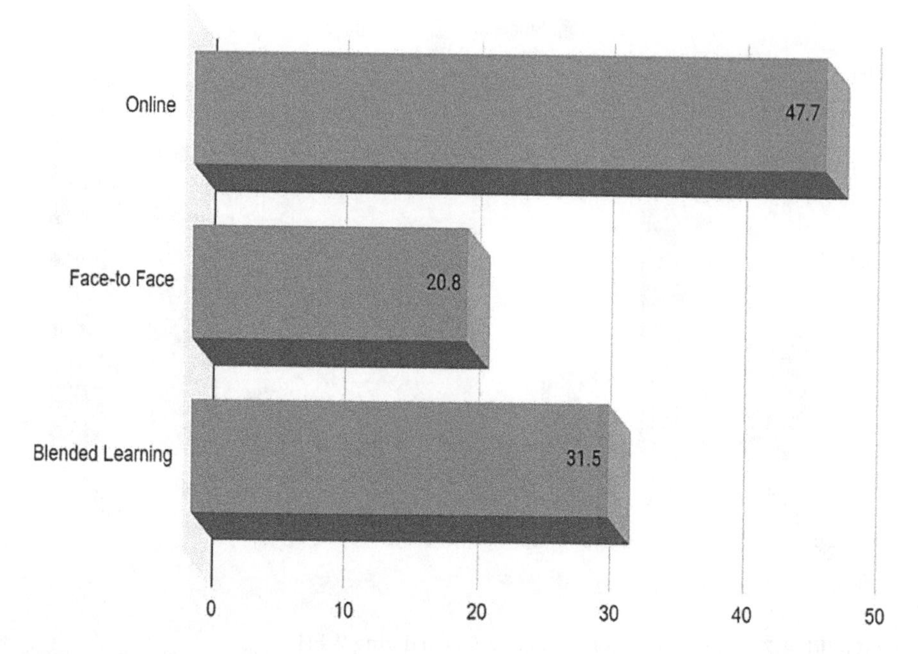

FIGURE 4.8 Future Shape of Institution.

4.6 CONCLUSION

The participant profile of this was the majority made up of males in the younger age group of 31 to 40. The majority held a postgraduate degree and were on permanent tenure. However, a small number had experience in academia for more than ten years. Despite this, the Professor rankings were the highest with respect to academic roles. There was a clear move from contact-based to online learning during the pandemic, with a slight increase in blended learning adoption. It was encouraging to note that despite not having any formal training on the adoption of online teaching and assessment tools, prior to the COVID-19 pandemic, academics are confident that their proficiency in technology tools to support online teaching and assessment is better than average. The most popular online teaching and assessment tools adopted by academics at this Institute during the pandemic included LMS tools like BlackBoard, virtual platforms like Zoom, and social tools like WhatsApp to support online teaching. There was also a marked increase in the adoption of Recorded Audio Lectures and Recorded Video Lectures to support their online teaching; while Google tools were utilized more during the pandemic for assessment than they were prior to the pandemic.

Furthermore, it was encouraging to note that the number of academics who do not adopt online teaching tools and/or online assessment tools dropped to a minimal number during the pandemic. Academics at this Institute felt that their students faced infrastructural barriers—such as access to the internet connection, lack of devices, as well as the social issue of lack of interaction during their online learning experience. Direct and personal methods of supporting students were most effective for academics at this institution—as they indicated that interaction during online lectures and online assessment, direct Emails to students, and personal phone calls were the best ways to support students during their online learning. Students were responsive to online learning and assessment, which led these academics to believe that online teaching was easy to work with and effective. Thus a significant number of respondents indicated that they would continue to adopt online teaching post the pandemic. Generally, internet connection for students and the academics themselves, as well as general anxiety about the COVID-19 pandemic, were the main challenges that academics at this institution faced with the shift to work from home. While social and LMS-based tools were adopted to some extent to communicate with students, the most popular tools adopted for communicating with students and colleagues in the WFH situation were Zoom, WhatsApp, and Email. It was clear that academics at this institution felt that the institution would be fully reshaped post-COVID-19 to adopting more online and blended learning approaches than face-to-face learning.

Data were collected during the final stages of the hard lockdown in India. This was a busy period for academics, having to transition into a fully online environment, in an attempt to ensure continuity of academic activities. While the results may prove interesting and provide insight during the pandemic, they cannot be projected onto the general academic population due to the low sample size.

The body of literature on the impact of COVID-19 on the education of senior classes and Institutions is still developing. Future research could focus on these academics' lived experiences in the transition to the online environment and its sustainability post the pandemic. It would be beneficial to investigate the impact of the transition to the online environment on students at this institution, as well as Administrative and Support staff at this institution.

BIBLIOGRAPHY

Allo, M.D.G. (2020). Is online learning good in the midst of COVID-19 pandemic? The case of EFL learners. *Journal Sinestesia*, *10*(1), 1–10. Available at: https://sinestesia.pustaka.my.id/journal/article/view/24. (Accessed on 11 May 2020)

Alruwais, N., Wills, G., & Wald, M. (2018). Advantages and challenges of using e-assessment. *International Journal of Information and Education Technology (IJIET)*, *8*(1), 34–37.

Appavoo, P., Sukon, K. S., Gokhool, A. C., & Gooria, V. (2018). Technology affordances at the open university of Mauritius. In *Technology for efficient learner support services in distance education* (pp. 153–171). Springer. doi:10.1007/978-981-13-2300-3_8

de Oliveira Araújo, F. J., de Lima, L. S. A., Cidade, P. I. M., Nobre, C. B., & Neto, M. L. R. (2020). Impact of SARS-cov-2 and its reverberation in global higher education and mental health. *Psychiatry Research*, *288*, 112977.

Bao, W. (2020). COVID-19 and online teaching in higher education: A case study of Peking University. *Human Behavior and Emerging Technologies*, *2*(2), 113–115.

Blackmom, S.J., & Major, C. (2012). Student experiences in online courses: A qualitative research synthesis. *The Quarterly Review of Distance Education*, *13*(2), 77–85.

Blewett, C., & Hugo, W. (2016). Actant affordances: A brief history of affordance theory and a Latourian extension for education technology research. *Critical Studies in Teaching and Learning*, *4*(1), 55–76.

Bothwell, E. (2020). *Flexible admissions could mitigate COVID-19 impact, Timeshighereducation.com.* Times Higher Education (THE). Available at: https://www.timeshighereducation.com/news/flexible-admissions-could-mitigate-COVID-19-impact (Accessed: 28 September 2020).

Butcher, N. (2014). *Technologies in higher education: Mapping the terrain.* UNESCO Institute for Information Technologies in Education. Retrieved from http://iite.unesco.org/pics/publications/en/files/3214737.pdf

Cape Argus. 2020. *This is how SA varsities are implementing online teaching amid COVID-19 lockdown.* Available at: 22https://www.iol.co.za/capeargus/news/this-is-how-sa-varsities-are-implementing-online-teachingamid-COVID-19-lockdown-46930129 (Accessed on 27 June 2020).

Cheng, R. (2020). *Higher ed institutions aren't supporting international students enough during the COVID-19 crisis (opinion), Insidehighered.com.* Available at: https://www.insidehighered.com/views/2020/03/19/higher-ed-institutions-arent-supporting-international-students-enough-during-COVID (c), 395–400.

Coronavirus COVID-19—latest update on Kingston University's response (2020) *Kingston.ac.uk.* Available at: https://www.kingston.ac.uk/news/article/2306/27-mar-2020-coronavirus-COVID19-latest-update-on-kingston-universitys-response (Accessed: 28 September 2020).

Crawford, J., Butler-Henderson, K., Rudolph, J., Malkawi, B., Glowatz, M., Burton, R., Magni, P. & Lam, S. (2020) COVID-19: 20 countries' higher education intra-period digital pedagogy responses, *Journal of Applied Learning & Teaching*, *3*(1). doi:10.37074/jalt.2020.3.1.7.

Current Funding Efforts. (2020) *Iie.org.* Available at: https://www.iie.org/Programs/Emergency-Student-Fund/Current-Funding-Efforts (Accessed: 28 September 2020).

Dill, E., Fischer, K., McMurtrie, B., & Supiano, B. (2020) *As Coronavirus Spreads, the Decision to Move Classes Online Is the First Step. What Comes Next?, Chronicle.com.* Available at: https://www.chronicle.com/article/as-coronavirus-spreads-the-decision-to-move-classes-online-is-the-first-step-what-comes-next/ (Accessed: 28 September 2020).

ET Government. (2020). *Post Corona: How we teach and evaluate must get overhauled, ETGovernment.* Available at: https://government.economictimes.indiatimes.com/news/governance/post-corona-how-we-teach-and-evaluate-must-get-overhauled/75117093 (Accessed: 28 September 2020).

Guo, Y.-R., Cao, Q.-D., Hong, Z.-S., Tan, Y.-Y., Chen, S.-D., Jin, H.-J., Tan, K.-S., Wang, D.-Y., & Yan, Y. (2020). The origin, transmission and clinical therapies on coronavirus disease 2019 (COVID-19) outbreak—an update on the status. *Military Medical Research*, *7*(1), 11.

Gupta, A., & Goplani, M. (2020). Impact of COVID-19 on educational institutions in India. Unpublished. doi: 10.13140/RG.2.2.32141.36321.

Gutterer, J. (2020). *The impact of COVID-19 on international students perceptions, Studyportals.com.* Available at: https://studyportals.com/blog/the-impact-of-COVID-19-on-international-students-perceptions/ (Accessed: 28 September 2020).

Houlden, S., & Veletsianos, G. (2020). Coronavirus pushes universities to switch to online classes—but are they ready. *The Conservation.*

Huelsman, M. (2020). *Coronavirus could have a long-term impact on state funding of universities (opinion), Insidehighered.com.* Available at: https://www.insidehighered.com/views/2020/03/12/coronavirus-could-have-long-term-impact-state-funding-universities-opinion (Accessed: 28 September 2020).

IAU. (2020). *The impact of COVID-19 on higher education worldwide Resources for Higher Education Institutions, Iau-aiu.net.* Available at: https://www.iau-aiu.net/IMG/pdf/COVID-19_and_he_resources.pdf (Accessed: 28 September 2020).

Jandrić, P., Knox, J., Besley, T., Ryberg, T., Suoranta, J., & Hayes, S. (2018). Postdigital science and education. *Educational Philosophy and Theory*, *50*(10), 893–899.

Kearns, L. R. (2012). Student assessment in online learning: Challenges and effective practices. *MERLOT Journal of Online Learning and Teaching*, 8(3). Available at: https://jolt.merlot.org/vol8no3/kearns_0912.pdf (Accessed: 28 September 2020).

Li, Y., Wu, S., Yao, Q., & Zhu, Y. (2013). Research on college students' online learning behavior. *e-Education Research*, 34(11), pp. 59–65.

Lim, M. (2020). *Educating despite the COVID-19 outbreak: Lessons from Singapore, Timeshighereducation.com.* Times Higher Education (THE). Available at: https://www.timeshighereducation.com/blog/educating-despite-COVID-19-outbreak-lessons-singapore (Accessed: 28 September 2020).

Mdepa, W. (2020). *I'm not against online teaching, but what about us?* Available at: https://www.dailymaverick.co.za/opinionista/2020-05-11-im-not-against-online-teaching-but-what-about-us/#gsc.tab=0 (Accessed on 26 June 2020).

Mantha, D. S. S. (2020) *Universities: Post CORONA, Businessworld.in.* Available at: http://www.businessworld.in/article/Universities-Post-CORONA/05-04-2020-188289 (Accessed: 28 September 2020).

Mansoor, S. (2020, March 12). India is the world's secondmost populous country. Can it handle the coronavirus outbreak? Time. https://time.com/5801507/coronavirusindia/

McKibbin, W., & Fernando, R. (2020). 'The global macroeconomic impacts of COVID-19: Seven scenarios', *Asian economic papers*, 1–55.

Mthethwa, A. (2020). *Remote learning challenges delay resumption of universities.* Available at: https://www.dailymaverick.co.za/article/2020-04-22-remote-learning-challenges-delay-resumption-of-universities/#gsc.tab=0 (Accessed on 26 June 2020.)

Mugo, K., Odera, N., & Wachira, M. (2020). *Surveying the impact of COVID-19 on Africa's higher education and research sector, Africaportal.org.* Available at: https://www.africaportal.org/features/surveying-impact-COVID-19-africas-higher-education-and-research-sectors (Accessed: 28 September 2020).

Mzileni, P. (2020). *How COVID-19 will affect students.* Available at: https://mg.co.za/education/2020-04-23-how-COVID-19-will-affect-students/ (Accessed on 28 June 2020.)

Pushkarev, S. (1975). AIU: Minimizing impact of COVID-19 on students. *Slavic Review*, *34*(4), 879–880. doi: 10.1017/s0037677900068716.

Queiros, D., & de Villiers, M. (2016). Online learning in a South African higher education institution: Determining the right connections for the student. *International Review of Research in Open and Distributed Learning: IRRODL*, *17*(5), 165–185.

Raaheim, A., Mathiassen, K., Moen, V., Lona, I., Gynnild, V., Bunæs, B. R., & Hasle, E. T. (2018). Digital assessment—how does it challenge local practices and national law? A Norwegian case study. *European Journal of Higher Education*, 9(2), 1–13.

Rumbley, L. E. (2020). *Coping with COVID-19: International higher education in Europe.* Available at: https://cbie.ca/wp-content/uploads/2020/04/EAIE-Coping-with-COVID-19-International-higher-education-in-Europe.pdf (Accessed: 28 September 2020).

Russell, F.R. 2019. *Adult learners: Standards for teacher effectiveness and conditions for optimal learning.* IGI.

Sahu, P. (2020). Closure of universities due to Coronavirus disease 2019 (COVID-19): Impact on education and mental health of students and academic staff. *Cureus*, *12*(4), e7541.

Salcedo, A., Yar, S., & Cherelus, G. (2020). Coronavirus travel restrictions and bans globally: Updating list. *The New York times*, 16 July. Available at: https://www.nytimes.com/article/coronavirus-travel-restrictions.html (Accessed: 28 September 2020).

Schwartz. (n.d.). Engaging adult learners. Ryerson University. Available at: https://www.ryerson.ca/content/dam/learning-teaching/teaching-resources/teach-a-course/engaging-adult-learners.pdf (Accessed on 25 June 2020).

Shang, J., & Cao, P. (2017). "Internet plus" and the reform of higher education: A preliminary study on the development strategy of higher education informatization in China. *Peking University Education Review*, *1*, 173–182.

Singh, U.G. (2020a). Academic digital literacy—A journey we all need to take. *University World News*. https://www.universityworldnews.com/post.php?story=20200630085507410. (Accessed 10th October 2020).

Singh, U.G. (2020b). *UKZN Academic Develops Framework for Online Learning—ASSET© (Academic SkillSET)*. YIBA. https://yiba.co.za/ukzn-academic-develops-framework-for-online-learning-asset-academic-skillset/ (Accessed 10th October 2020).

Singh, U.G., Proches, C.N., Leask, C, & Blewett C.N. (2020). Emergency Online Learning (EOL) during the COVID-19 pandemic: Postgraduate students' perspectives. digiTAL 2020 Conference Proceedings, Durban, South Africa.

Splashgain. (2018). *Top 7 trends of education technology for 2020—education technology for digital assessments, exams, admissions and trends, Epravesh.com*. Available at: https://www.blog.epravesh.com/top-7-trends-of-education-technology-for-2019/ (Accessed: 28 September 2020).

Sujatha R, & Chatterjee J. (2020). *A machine learning methodology for forecasting of the COVID-19 cases in India.*

Tejaswi, D. (2020). *Can we turn this crisis into an opportunity?, Highereducationplus.com*. Available at: https://highereducationplus.com/can-we-turn-this-crisis-into-an-opportunity (Accessed: 28 September 2020).

The Economist. (2020, March 21). The billion-person question if COVID-19 takes hold in India the toll will be grim. *The Economist*. https://www.economist.com/asia/2020/03/21/if-COVID-19-takes-hold-in-india-the-toll-will-be-grim.

Timmis, S., Broadfoot, P., Sutherland, R., & Oldfield, A. (2016). Rethinking assessment in a digital age: opportunities, challenges and risks. *British Educational Research Journal*, *42*(3), 454–476.

TNN. (2020, March 17). COVID-19 scare: Pondicherry University shuts down. *The Times of India*. https://timesofindia.indiatimes.com/city/puducherry/COVID19-effect-pondicherry-university-shuts-down/articleshow/74663011.cms

UGC. (2020a). *Redressal of Grievances Related to COVID-19 pandemic, Ugc.ac.in*. Available at: https://www.ugc.ac.in/pdfnews/5062771_Public-Notice---Redressal-of-Grievances-Related-to-COVID-19-Pan.pdf (Accessed: 28 September 2020).

UGC. (2020b). *UGC Guidelines on Examinations and Academic Calendar for the Universities in View of COVID-19 pandemic and Subsequent Lockdown, Ugc.ac.in*. Available at: https://www.ugc.ac.in/pdfnews/4276446_UGC-Guidelines-on-Examinations-and-Academic-Calendar.pdf (Accessed: 28 September 2020).

USAF. (2020). *Online learning is integral to the future of Higher Education; embrace it or become irrelevant*. Available at: https://www.newssite.co.za/usaf/online-learning.html (Accessed on 27 June 2020.)

Wang, W., Guo, L., He, L., & Y.J. Wu. (2019). Effects of social-interactive engagement on the dropout ratio in online learning: insights from MOOC. *Behaviour & Information Technology*, *38*(6), 621–636. Available at doi:10.1080/0144929X.2018.1549595. (Accessed on 5th May 2020.)

Watson, G., & Sottile, J. (2010). Cheating in the digital age: Do students cheat more in online courses?. *Online Journal of Distance Learning Administration*, *13*(1). Available at: http://www.westga.edu/~distance/ojdla/spring131/watson131.html (Accessed: 28 September 2020).

Wenham, C., Smith, J., Morgan, R., & Gender and COVID-19 Working Group. (2020). COVID-19: The gendered impacts of the outbreak', *Lancet*, *395*(10227), 846–848.

Wilson, S.E. (2018). Implementing co-creation and multiple intelligence practices to transform the classroom experience. *Contemporary Issues in Education Research*, *11*(4), 127–132. Available at doi:10.19030/cier.v11i4.10206 (Accessed on 20 March 2020).

Worldometers. (2020). *Coronavirus update (live): 33,360,594 cases and 1,003,174 deaths from COVID-19 virus pandemic—worldometer, Worldometers.info*. Available at: https://www.worldometers.info/coronavirus/ (Accessed: 28 September 2020).

Xudong, Z. (2020). It is urgent to build a national online teacher education system. *Educational Development Research*. doi:10.14121/j.cnki.1008-3855.2020.02.002.

Yamin, M. (2020). Counting the cost of COVID-19. *International Journal of Information Technology*, *12*(2), 311–317. Available at https://:10.1007/s41870-020-00466-0. (Accessed on 10 June).

Zhu, X., & Liu, J. (2020). Education in and after COVID-19: Immediate responses and long-term visions. *Postdigital Science and Education*, doi:10.1007/s42438-020-00126-3.

5 Adaptability of Computer-Based Assessment Among Engineering Students at Higher Educational Institutions

Muhammad Mujtaba Asad, Ali Muhammad Mahar, Al-Karim Datoo, and Zahid Husaain Khand
Sukkur IBA University, Pakistan

Amir A. Abdulmuhsin
University of Mosul, Iraq

CONTENTS

DOI: 10.1201/9781003102298-5

5.1 BRIEF HISTORY OF COMPUTER-BASED ASSESSMENT

In the mid of the year 1985, a college board introduced a computer-based test (CBT), which was composed of four tests: counting, arithmetic sentence skills, reading comprehension, and basic polynomial math. All four tests were kept at low stake because they were designed to assist students in English and Science courses in a college. The very high-stakes examination directed via CBTs was Novell Corporation's certified network engineer (CNE) examination. Concurring to Miri 2016, the CNE was conducted online at Drake Prometric testing centers in 1990 and changed to online CAT in 1991. The Graduate Record Examinations (GRE) an education testing service (ETS) examination persuade CNE which was functionally organized as a CAT at Sylvan testing centers over the US (Mills & Stocking, 1996). In 1994, at a commercial testing center, employing a CAT, two NCLEX examinations for nursing candidates were actualized (Zara, 1994). A CAT form of the Armed Services Vocational Aptitude Battery (ASVAB) was conducted online at Military Entrance Processing Stations (Sands, Waters, & McBride, 1997). Besides, in 1997, GMAT tests were taken too via CAT. Also, the Architect Registration Examination (ARE) was rolled out in 1997, advertising intelligently, computer-aided building issues inside a custom graphical interface. The architect examination was taken after by the US Medical Licensing Examination (USMLE) modulating to CBT in 1999. In this examination, exceedingly intuitively computerized patient management reenactments to the examination were blended (Clyman, Melnick, & Clauser, 1995). Intelligently bookkeeping reenactments were included in the uniform CPA examination in 2004 (Devore, 2002). And executed one of the primary computer-adaptive multistage testing systems for large-scale applications (Breithaupt, Ariel, & Hare, 2010). The successive flow of the world change toward 21st century has run through around the world. Within the past decade of the 21st century, assessment has ended up the most attentive subject in the educational estimation community (Geisinger, 2016). There has been a modern educational worldview called online learning and assessment, it has got a longer preference for the teaching of higher education in Malaysia. Because of headway in the field of innovation and technology candidates have the choice to decide the place and also to enjoy the freedom of learning anything at any time, i.e. when and where they need to learn (Mahat et al., 2012). This world is changed in each course. Humans have evolved themselves into a modern period from the dull ages through insights and intelligence and another such kind of abilities. As a result, man has created such devices that are exceptionally useful and valuable to influence human life less demanding and more sumptuous. A startling transformation has been there within the field of innovation and technology, i.e. alteration in communication to transportation to amusement. In a plain word, computer and other devices have been introduced that have reserved higher significance in one's day-to-day routine works. The globalization and internalization of the economy and the fast advancement of information and communication technologies (ICTs) are ceaselessly renewing how we live, work, and learn (Voogt & Roblin, 2012). These days with the favors of technology and technological tools, things are at our fingertips, and we can view each modest development on the globe. Adding to this, technology plays an outstanding part in improving the educational dimensions and looking at instructions with a more profound focal point. Greenstein (2012), in her book, expressed that one of the central abilities required in 21st century is applying and coordinating information and communication advances in education. In our context, there is a lack of digitally literate teachers and educators who are unable to operate technological tools helpful for teaching and learning processes inside the classrooms. Likewise, our students are unaware of the knowledge and implications of such tools in the field of education.

Therefore, to meet this skill of applying and integrating ICTs in education, educators and stakeholders have to design a framework that must be based upon the curriculum in which there is a focus on technology and technological tools up to a great extent. By integrating ICTs in our curriculum, we can produce digitally literate teachers, educators, and graduates which is the need of the hour. It can be obvious that in society the eruption of technology has constrained amendments in work and education (Fazilat Siddiq, & Perman Gochyyev, 2017). If we will have a glance around in the job market, the needs of employees have been transformed at a wider level. The recruiting companies are hiring digitally literate graduates. Moreover, the learning and teaching process also takes place in the classroom as a result of it. There are projectors, multimedia tools, tablets, PCs, laptops, and smart mobile phones in the hands of students. You will hardly find paper, notebook, textbook, pen, pencil, and any traditional material in the classroom. Certainly, it has enhanced the process of teaching and learning. Similarly, computers and other electronic tools help assess students' learning and performance. Quellmalz (2014) included a section in a chapter in the education encyclopedia which talked about assessments in the next generation of science standards, where science phenomena need more flexible, dynamic, and more complex representation. The field of science is very vast and uncovering Science subject needs more dynamic methods to know, understand, and make inventions and discoveries. For this, assessment techniques need to be changed. Assessment tools must be designed in a way so that educators can easily express their ideas, can apply their cognitive abilities in a real scenario, and can work for the betterment of students.

5.2 ASSESSMENT APPROACHES

The key factor of all education system is assessment as it has main role in building a career of students because it is a dilemma in our context that we measure students with grades which they get after being assessed. It has the key function in overall education and teaching and learning process revolves around it. It is not only effective for students but also for teachers and administrators as it facilitates students, helps teachers in improving instructions, and guides administrators in deciding the process of resource allocations. At upper level, it is also beneficiary for policy makers because through assessment they evaluate the efficiency or efficacy of any educational program. Moreover, it provides students and teachers a path toward their improvement. One gets satisfaction and can be unsatisfied due to assessment because we get results through it in order to know our place or position in the classroom through measuring our academic performance. Teaching and learning process can take place at anytime and anywhere due to development of web browsers and technological tools which has enabled the use of internet for educators, students, and teachers over the globe. It is important to note that assessment is not an easy job to do as it requires basic understanding of it and requires an in-depth knowledge to be an effective expert of the field. Besides this, it is time-taking and complex process where students' knowledge is being assessed. We can overcome those challenges by introducing the computer-based assessment (CBA) and taking full advantage of this new technology in the field of education (Cook, 2007). In order to make ease in the process of assessment, this type of technological tool online assessment (CBA) system has been successfully implemented in various domains (Thompson, 2008).

5.3 INTEGRATION OF TECHNOLOGY IN ASSESSMENT

With the emergence and progress of the ICT, CBA is considered as an efficient, innovative, and evolutionary assessment method in educational environments to evaluate students' performance compared with conventional paper-based one. As a result of invention of internet, there is no border to any field especially education and educational assessment. It has made life easier for students and teachers to assign, submit, assess, and evaluate the work at any place, anytime. As compared to traditional practices of assessment, CBA saves time and resources. Education has become available at the globe and anyone can get education at one's doorsteps without any fluctuations regardless of any limitations, i.e. time and place. For students and educators, it has become more convenient to attend and listen to lectures of world's famous teachers and leaders by sitting in their classroom and offices, respectively. Due to being economical, CBA is preferred over paper-based assessment (PBA) as it is advantageous in terms of time, money, and other resources. In addition, paper and pencil tests are being outdated slowly and gradually as assessments are now increasingly being designed as adaptive and delivered online (e.g., computer adaptive testing), employing dynamic and interactive tasks and simulations (Luecht, 2013). Before CBA, roles played in the classroom were different but now with its emergence and wide adaptability, it has shifted the roles of teachers and students in the classroom (Gray et al., 2004; Volman, 2005). As in teacher-centered classroom, he is the one who leads learning and is an authoritative person and center of the learning. On the other hand, e-learning puts this reversely by emphasizing more on students rather than teachers. It means student is the main center of learning and leads the classroom activities. In e-learning classroom environment, teacher is only a facilitator of learning and plays the role of a guide for learning and instructor for learning. For students, they are the key players of the learning in the classroom as they are the drivers of the classroom in e-learning environment. In such classrooms, there are students who decide what to learn, how to learn, and when to learn. Therefore, in order to meet the standards of current educational system in the world, we have to consider the interests and learning styles of the students.

5.4 COMPLIANCE OF COMPUTER-BASED ASSESSMENT

Modern world has brought so many technological advancements in the field of education, especially in assessment so as to meet the national standards, and can compete at higher educational institutes in the globe. The time has changed to produce graduates with minimal skills as now educational institutes have to produce graduates with the 21st-century skills, i.e. cooperation and collaboration, creativity, communication, critical thinking, and digital literacy, etc. Due to development in technological tools and their integration in assessment activities, teachers can evaluate various, such as cognitive, affective and metacognitive, aspects of their children (Pellegrino et al., 2001; Raymond & Usherwood, 2013). According to Geisinger (2016), there are four types of 21st-century skills: collaborative problem-solving, complex problem-solving, creativity, and digital information literacy. Contrary to PBA's classroom settings, it is important to create such environment in the classroom where students can learn through digital means so as to improve their familiarity with digital tools. To ensure this practice in the educational institutes, educators have to shift from traditional practices to interactive methodologies and approaches in the classroom. It is not enough to transform teaching and learning practices only but with this we have to bring amendments in our overall examination system because without changing it, other practices will not be beneficial for us as much as expected. As we know that our examination system is static and it is not helpful at the present time to conform to this modern world. Question papers must be designed as per classroom instructions and practices considering the skills for 21st century. Classroom teaching should be interactive so that students can work collaboratively with each other. Moreover, teachers must put such problems which require students to think deeply, undergo the process, and bring the solution to it. We are aware of the examination system that determines the performance, learning, and abilities of students.

5.5 21ST-CENTURY TEACHING AND ASSESSMENT PRACTICES

Educational assessment is a fundamental aspect of a teaching and learning process. The purpose of this assessment, as a systematic process, is to collect, analyze, interpret, and evaluate relevant information about students' progress and make judgments in order to achieve the educational objectives. In 21st-century learning and assessment goes beyond the basics of reading, writing, interpretation, and synthesis; but students need to ensure that they also have the ability in mastering some of the skills: cognitive, interpersonal, intrapersonal, and technical skills hereafter called OECD approach (Geisinger, 2016). It is also important to apply different assessment techniques rather than relying on old traditional methods of assessment. Assessment is key to learning design (Armellini & Aiyegbayo, 2010). In early times, assessment was carried only through pen and paper system. Nowadays, educators have designed latest techniques of assessment in the field of education. Examination system leaves a great impact on students' learning process and teachers' teaching process, Rehmani (2003). Educational assessment has shifted from paper and pen to digital assessment through computers and other electronic tools. Previously, relying on paper and pen system, teachers were unable to assess the complex form of knowledge and reasoning because it is impossible to assess higher order learning through paper and pen system. According to Bodmann and Robinson (2004), technology-based assessment (TBA) opens door for assessing students' complex form of knowledge, reasoning, and other higher levels of cognition. Traditional assessment does not help students in improving higher levels of cognition. On the other hand, TBA provides students a platform where they can respond through multiple means, i.e. pictorial, graphical, text, audio, video, simulations, and animations.

5.6 ONLINE LEARNING AND ASSESSMENT

There has been a great transformation in the process of teaching and learning since the use of online learning and assessment. With its emergence and wide adaptability at higher educational institutions, students and teachers have kept themselves away from the burden of heavy books and notes for record keeping and maintaining. Moreover, it is more advantageous than PBA in a way as it is environmentally friendly since it does not need paper and pencil; rather, it uses technological tools to conduct examinations. It is also economical for students as they do not need bunch of papers for printing and submitting their assignments and projects for assessment. The terminology widely used in online learning and assessment is mobile learning (M-learning), E-learning, and Kahoot that are new tools of learning and assessment via online mode. According to Mahat, Fauzi, Ayub, and Luan (2012), information can be transferred through multiple modes in M-learning environment such as laptops, mobiles, and computers. Students' activeness has increased as a result of using their mobiles for learning and assessment purposes as they can participate in classroom discussion, share material, and submit assignments through M-learning. Research found that mobile learning promotes the elements of collaborative work and cooperation among individuals, and it is also helpful in extrinsic motivation (Alvarado, Coelho, & Dougherty, 2016). In 21st century, educators and teachers do not need to invest much time on teaching and training students in order to make them familiar with technological tools as this generation is already technology savvy. One of the core skills of this century is to master technological tools and their usability.

5.7 ADVANTAGES OF COMPUTER-BASED ASSESSMENT

CBA offers the following advantages against traditional paper-based examination:

- Grades and feedback can be immediately provided to the students.
- Greater flexibility in the location and time to carry out the exam is allowed.
- Interactive activities and multimedia tools can be incorporated to the assessment process.
- Teacher's interpretations and legibility problems are avoided.
- Time and cost effectiveness, among others, for even the largest class sizes—less time is needed for marking.
- Greater frequency of assessment is allowed, providing detailed information on the students' progress and helping teachers to early detect possible problems.
- Motivation of the students is increased and skills in the use of ICTs are encouraged.

Recent and previous evidence in journals and publications show that online assessment offered several benefits to educators, students, invigilators, and faculty (Hertel, Naumann, Konradt, & Batinic, 2002; Lim et al., 2006). The finding in Hoskins and

Van Hooff (2005) shows that the achievement of students via an online assessment learning platform is extremely promising.

1. **Auto-Marking**: After conduction of examination assessment/grading is the primary duty but teachers hardly find time to assess students' work. This issue can be addressed by educators by themselves through utilizing online platforms in order to grade students' work or they can also develop their own by creating their own way. Without making students wait so much for the results, educators can easily mark their work and can issue the results right after students have done with their examination. Auto-marking was thus possible, but answers had to be submitted a few days in advance to the computer center for programming purposes (Lim et al., 2006). This is the best way to save time and very reliable as students could not alter the answer and also to avoid human error when marking the questions.
2. **Quality Feedback**: Eom et al. (2006) took into account the importance of feedback in the student learning process and student satisfaction. Feedback is one of the most important components in the assessment. Effective assessment is when online educators can adapt their assessment activities to provide useful feedback, accountability, and opportunities to demonstrate quality (Robles & Braathen, 2002). With the help of quality feedback from educators and students, educators themselves can manage their students' performance and set a target for each student throughout the year. As in Vonderwell and Boboc (2013), the feedback in formative assessment can foster student engagement, improve achievement, and enhance motivation to learn. There are two main aims of computer-based assessment for learning (CBAfL) in classroom: (a) provide appropriate and timely feedback to students and (b) personalize learning. Improving the quality of feedback and the way it is delivered are critically important components of learning (Farrell & Rushby, 2015; Shute, 2008; Timmis et al., 2015; van der Kleij, Eggen, Timmers & Veldkamp, 2012), and computers can help to accomplish these two goals (Thelwall, 2000).
3. **Reliable and valid measurement**: In today's assessment reliability and validity matter a lot. The question and marking in online setting is reliable and valid. To establish a valid and reliable CBT, the International Guidelines on Computer-Based Testing (International Test Commission 2004) stated that equivalent test scores should be established for the conventional paper-based testing (PBT) and its computer-based mode (Piaw, 2012). Firstly, it allowed faculty to make online changes to the questions and answers prior to publication of the examination; and secondly, there was little chance of a security breach, as only the "webmaster" in charge of the examination would have access to the paper (Lim et al., 2006).
4. **Economic and Ecological**: The nature of using online setting itself is environmentally friendly as it is paperless. Online assessment platform can be convenient, cost-effective, and environmentally friendly. Expenses in conducting the online assessments are very low because time and materials can be minimized and all the data acquisition and analyses can run automatically. Tons of paper would be saved within a year if schools, universities, and educational

institutions were to replace paper–pencil tests with CBTs (Piaw Chua, 2012). Reducing paper consumption will indirectly reduce greenhouse gases and energy consumption.

5. **Practical**: In this 21st century, formative e-assessment is seen as having the potential to support significant changes in the way in which learning occurs in Higher Education (Pachler, Daly, Mor, & Mellar, 2010). Online assessment is practical. It can be done anytime and as what is planned by educators. It also enables a more flexible pace of learning. Participants respond faster than paper and pencil questionnaires. Online learning is providing higher education institutions with an entirely new modality for educating learners free from the constraints of time and location (Albee, 2015). It does not need to be printed before conduct of the test. It discourages cheating through randomizing the questions. Results of the test can be provided immediately soon after the test has been taken. It is easy to maintain the profile of students, bundle of questions, and past papers and results. It is time-saving assessment as large number of students' responses can be assessed simultaneously. The possibility to examine a student several times and create adaptive tests, based on the pattern of wrong answers given to previous tests.

There are many online learning resources such as YouTube channels. For an average school or academy, the monthly fee is not less than 5000, but for A-level there is about 40,000–50,000 expense per month. Therefore, online education is better than face-to-face learning as it saves a lot of money. The advantage of online education is at low or no expense, vast volumes of prepared material are available, most of which are digital (Jahangir, 2020). Due to tensions among countries, there is possibility in future that students may be unable to go abroad for higher studies, so, online learning will fulfill their need of getting quality education at doorsteps. It is also the charm of online education that it helps in removing social, physical limits and obstacles as in rural areas girls are probably not allowed to go for higher studies in other cities. In order to resolve this issue, online education is the best way to ensure high-quality education regardless of investing on schools and resources (Lau, 2020). Through the trend of online education, there will be great revolution in job market. Skilled educators and trainers will widen their scope through online education. As we are lacking in many areas including quality education, so, Pakistani students will be able to learn from world-known professors in online classes. Since decades there has been limited development in the education sector, especially in Pakistan, so, it is an opportunity for educators and policy makers to transform our educational practices and policies. In this way, we can keep ourselves up to date in order to compete at global level. In early times, teaching and learning processes were only taking place through books, chalk, and blackboard. However, students are now only a finger click away from massive knowledge present at different websites which is much better than actual teachers' knowledge transferred in the classroom. Technology has transformed the mobile screen into an interactive platform where students and teachers can discuss, explain, and debate regardless of presence in the classroom. In the future, virtual classrooms could allow students to attend school in person for, say, four days, with fifth-day online lessons (Baker, 2020). Students and teachers are now learning how to use technological tools for schooling purpose. They are using emails for communication, LMS for discussion and resource sharing, and video conferences for delivery of lessons. Moreover, students have created different groups for discussion and educational talks on different platforms, i.e. WhatsApp and Facebook. The use of paper and pen is negligible in learning and teaching. In traditional classroom, there was a limitation over number of students in each class but in online classes this limitation is no more there and any of registered students in course can attend class virtually. In simple words, the large number of students in online classes does not disturb teaching and learning process as in face-to-face classes. Keep in mind, online learning, of course, does not mean that there should be no class size limitations at all, since it would remain necessary for students to connect with professors, teaching assistants, and their social group. But for most classes, without greatly altering the student's college experience, attendance could be increased by 10–20%. Online information can be revisited and accessed at any time, so, the concept of paying attention is traditional or outdated nowadays. Students used to forget major points in face-to-face classes whenever they lost their attention, but now this is not the case as they are able to go back and recheck the concepts for better understanding.

5.8 STUDENTS' PERCEPTIONS ON COMPUTER-BASED ASSESSMENT

The very crucial activity of an academic learning process is the assessment through which students are evaluated in terms of their knowledge and overall performance. As in different educational institutes over the globe, multiple strategies or methods of assessment are being used to evaluate students' performance. It is observed in our context (Pakistan) that assessment only takes place in two forms, i.e. either through PBA or CBA. But, CBA is hardly used in our educational institutes although we are surrounded by technological tools. Previously, due to number of reasons there was an emphasis put up on PBA but in present times, as a result of burst in technological world, educational institutions are shifting from traditional methods of assessment to CBA or e-assessment (Dube, Zhao, and Ma, 2009). Nikou and Economides (2013) stated that assessment plays an important role in pedagogy as it can be used by teachers as an instrument to know about their students' performance. Besides this, it is also helpful for students to get motivation in order to perform better in the course of their learning (Cox, Imrie, Miller and Miller, 2014). Generally, great number of students in our context prefer PBA as it is the only way being used to assess students' performance in our schools, colleges, universities, and standardized testing. Not only students but teachers are also reluctant to move themselves to CBA because they do not want to come out of their comfort zone. Over the years, in our education system, assessment is done through the use of paper and pen, (Demirci, 2007). On the other hand, the advantage teachers can get through CBA is that they can improve their pedagogical skills by reflecting on students' performance. According to Rollings-Carter (2010, May 28), such assessment taking place through the use of paper and pen is known as PBA. There are various factors that lead to preferring and using of paper and pen assessment, such as; lack of technological tools and awareness, teachers lacking ICT skills, readiness from both sides, i.e. teachers and students, nature of examination, and traditional mindset. Nevertheless, new methods of assessment have been developed over a period of time due to advancement and development of ICT. Before this, ICT was only used in education for the purpose of learning and teaching process (Russell and Haney, 2000). But now new mode of students' assessment has been implemented by educational institutes as a result of technological revolution (Gipps 2005). This new mode of assessment has allowed us to use computer and other technological tools, i.e. projectors for playing videos, delivering lectures, images, animations, and PowerPoint to make teaching and learning process effective. We were using computers and other tools for only classroom activities and were ignoring its efficiency in other areas, i.e. assessment. In present times, at international level, there has been great use of computers in schools, colleges, and universities for primary purpose of assessment. Bull and McKenna, (Bull & McKenna, 2000) referred to this new type of assessment as CBA.

In such assessment, students have the option to choose the way they want to represent their responses through computers, mobile phones, and laptops. It allows teachers to design question papers through which students can be assessed by every angle. In this way, teachers can assess students' creativity and critical thinking skills which are essential for 21st century. Apart from this, CBA also lessens the burden from teachers' shoulders as PBA is time taking. In case the assessment is paper based and the number of students is too large then it requires teacher's great amount of time to assess them. The CBA is more advantageous than PBA (Nikou and Economides, 2013). With (CBA), multimedia and graphics can now be used into assessment questions (Segall, Doolen, and Porter, 2005). Moreover, different types of questions add value to the question paper as well as in responses. It provides students and teachers a wide range of variety, opens their mind, and provides a deeper understanding of the concepts. Terzis and Economides (2011) listed some positive aspects of CBA: keeping records automatically, generating results very fast, time saving, and economic friendly. Contrary, PBA needs a lot of resources to conduct and assess a test as compared to CBA. Moreover, CBA has the efficiency to create a random set of questions from a large set of questions, automatic scoring and generating assessment reports are other benefits of CBA (Nikou and Economides, 2013). Besides this, (Chua, 2012) stated that reduction in cost associated with paper consumption as another advantage of CBA. In traditional PBA, schools have to invest a lot of money for resources to conduct examination. There are thousands of students; therefore, thousands of papers are needed to assess those students. DeRosa (2007) further stated that as paper consumption is reduced due to CBA, it leads to an indirect reduction of greenhouse gases and consumption of energy. As papers are made from trees, therefore, huge use of PBA has also resulted in cutting a great number of trees which directly affects our environment. Such benefits have forced academic and non-academic institutions to shift from PBA to CBA (Mills, Potenza, Fremer, & Ward 2005). As a result of offering ample benefits, CBA has an increased demand, especially in educational institutes (Triantafillou, Georgiadou, & Economides, 2008). There is no doubt that CBA consumes less expense than PBA; therefore, those increased expenses on PBA can be used on other areas of improvement within any educational organization. CBA has earned its name not only up to school level, but it is also widely used in high-stakes testing programs, entrance into tertiary institutions, and job promotion exercises (McFadden, Marsh, & Price, 2001). Additionally, it is used in standardized testing such as "TOEFL (Test of English as a Foreign Language)" and "GRE" (Jeong, 2012).

5.9 RELATIONSHIP BETWEEN STUDENT ASSESSMENT METHOD AND E-LEARNING SATISFACTION

In the literature on e-learning user satisfaction (e.g., Ho and Dzeng, 2010; González-Gómezetal et al., 2012), some research works have taken into account the assessment method as an explanatory factor of user satisfaction (Sun et al., 2008; Kelly et al., 2007). In relation to the different assessment activities, the very first thing is sitting for several hours for an exam puts pressure and stress on the student. It takes enough time for students to wait for start of the test and till that they are under pressure and doing nothing but worrying about and overthinking about their grades. A wide range of research papers have been published on this topic and it has been measured throughout the research (Cassady & Finch, 2015; Hoferichter et al., 2015). It is clear that one cannot perform well when he is stressed. In addition to this, there is evidence that stress negatively affects student's performance (Neemati et al., 2014; Crisan & Copaci, 2015).

Students observe that there is difference between face-to-face learning and online learning (e-learning); therefore, it is mandatory to have a specific analysis between the relation of students' satisfaction and assessment method (Paechter & Maier, 2010). Some students think that they learn better when they are given the chance to interact in a physical classroom. While others think that they can perform better if provided the chance to learn through e-learning and assessed by the same type of assessment (CBA). Several studies have been conducted on the effectiveness of CBA compared to PBA in relation to students' learning motivation (Schmeeckle, 2003; Chua, 2012; Nikou & Economides, 2016). Furthermore, Lemos & Nueza (2012) explored the relationship in their studies between e-learning students' expectations and their level of satisfaction. They considered several dimensions: course design, coordination, faculty and tutors; curricular program; resources learning methodologies, evaluation system, support services, and technological infrastructures. Those all elements were considered while determining the students' satisfaction level and expectations of the e-learning. Besides this, there is another tool called clarity through which we can determine students' satisfaction with the course (Paechter & Maier, 2010). It is important for students to observe that course material is effective for their learning (Diep et al., 2017). Therefore, students should go through the course outline before registering into the course so as to ensure their interest for learning. Students' satisfaction level does not only depend up on course material, but it is also aligned with the assessment activities. So, these activities must be aligned with course goals in order to increase the satisfaction level of students. In short, the only way to increase the students' satisfaction level with e-learning, there must be interactive activities which promote discussion between teacher and students in the classroom (Swan, 2001; Martín-Rodríguez et al., 2015).

5.10 EFFECTS OF CBA ON STUDENTS' PERFORMANCE

Assessment is the fundamental aspect of education which is inseparable to make teaching and learning process effective. It is the only way through which students are satisfied as they get to know about their position in terms of learning. Although there are various types of assessment but summative and formative assessment (continuous assessment) are of major use in education. There are multiple modes of assessment, i.e. PBA and CBA. Living in this world of technology, an efficient way to conduct continuous assessment is using online multiple-choice questions (also called quizzes), as they are relatively easy to implement (Martí Ballester & Orgaz Guerrero, 2014). Using online quizzes are more suitable and easy to conduct any test when there is huge number of students in the classroom. Electronic devices are widely used in our daily life routine and are given much importance in the learning process (Alvarez, Brown, & Nussbaum, 2011). Those devices are incorporated everywhere, from schools to universities, from homes to shopping, and from paying utility bills to shopping grocery. Furthermore, Islamic teachings are given through electronic devices especially tablets and mobile phones up to greater extent in religious schools (Madrasahs). In classroom, electronic devices are widely used in every classroom activity, either it is teacher or student, both are using electronic and technology tools so as to improve teaching and learning process. In recent decades, online multiple-choice questions have been increasingly used in higher education to assess students' knowledge. There are a number of factors which have motivated the increasing use of online quizzes as learning resources, such as the growing number of students, decreasing resources in the universities of several countries, and increased usage of computers, tablet PCs, and mobile phones (Nicol, 2007). Most of the standardized tests (for admission and employment) are based on multiple-choice questions; therefore, in order to be successful in these tests, one has to be familiar with use of such devices. Students are more satisfied with online multiple-choice questions, compared to paper-and-pencil quizzes, as they need less time to answer online questions (Segall et al., 2005). Online assessment allows teachers and students to spend less time in designing questions and attempting paper, respectively, and devote more time to teaching and learning in the classroom. It will also increase students' performance in tests as they will have more time for preparation. It is found in previous literature that using tablets leaves a positive impact on students' performance (Enriquez, 2010; Huang & Lin, 2017) and that students expect the use of tablets to perform tasks to improve their learning processes (Ifenthaler & Schweinbenz, 2016). However, it is evident that these positive effects depend upon cultural and socio-economic background of the student (Ferrer et al., 2011), as well as their technological literacy and learning style, especially in underdeveloped areas (Pruet, Ang, & Farzin, 2016).

5.11 CONCLUSION

It is evident from the literature that CBA is more advantageous than PBA, commonly known as traditional assessment. Although, there are some key challenges for the advocates of CBA, but these challenges possess less weightage in front of advantages it offers. Moreover, these challenges can be overcome by proper training and acquiring digital literacy. CBA is beneficial for both, i.e. students and teachers as it helps them to unlearn the previous assessment type (PBA) and relearn this new type of assessment (CBA) so as to be an effective learner and teacher, respectively. Many institutions are shifting toward CBA as the concept of paper and pen assessment has been outdated. It maximizes students' academic performance and teachers' pedagogical skills by reflecting on students' academic performance. In short, previously technology was being used only in teaching and learning process, but now it is carried out in assessment process as well.

REFERENCES

Albee, B. L. (2015). *Technology use of online instructors with high self-efficacy: A multiple case study.*

Bodmann, S. M., & Robinson, D. H. (2004). Speed and performance differences among computer-based and paper-pencil tests. *Journal of Educational Computing Research, 31*(1), 51–60. DOI: 10.13140/RG.2.2.32040.88326 https://www.researchgate.net/publication/324208499

Breithaupt, K., Ariel, A. A., & Hare, D. R. (2010). Assembling an inventory of multistage adaptive testing systems. In W. van der Linden & C. Glas (Eds.), *Elements of adaptive testing* (pp. 247–268). Springer.

Bull J., & McKenna C. (2000). Computer-assisted assessment centre (TLTP3) update, In: M. Danson (Ed.) *4th International CAA Conference, Loughborough University*, 21–22 June 2000. ISSN: 0968-7769 (Print) 1741-1629 (Online) Journal homepage: https://www.tandfonline.com/loi/zrlt19. https://doi.org/10.1080/0968776042000339772.

Cassady, J.C., & Finch, W.H. (2015), Using factor mixture modeling to identify dimensions of cognitive test anxiety. *Learning and Individual Differences, 41*(2015), 14–20.

Clyman, S. G., Melnick, D. E., & Clauser, B. E. (1995). Computer-based case simulations. In E. L. Mancall & Ph. G. Bashook (Eds.), *Assessing clinical reasoning: the oral examination and alternative methods* (139–149). American Board of Medical Specialties.

Cook, D. A. (2007). Web-based learning: Pros, cons and controversies. *Clinical Medicine*; *7*, 37–42.

Cox K., Imrie B. W., Miller A., & Miller A. (2014). *Student assessment in higher education: a handbook for assessing performance*. Routledge.

Crisan, C., & Copaci, I. (2015), The relationship between primary school childrens' test anxiety and academic performance, *Procedia-Social and Behavioral Sciences, 180*(2015), 1584–1589.

Demirci N. (2007). University students' perceptions of web-based vs. paper-based homework in a general physics course, *Online Submission, 3*, 29–34.

DeRosa J. (2007). *Reducing greenhouse gas emissions one ream at a time*. Global Warming Initiatives Inc.

Devore, R. (2002). Considerations in the development of accounting simulations. Paper presented at the Annual Meeting of the National Council on Measurement in Education, New Orleans, LA.

Diep, A.N., Zhu, C., Struyven, K.andBlieck, Y. (2017), Who or what contributes to student satisfaction in different blended learning modalities? *British Journal of Educational Technology, 48*(2), 473–489.

Eom, S.B., Wen, H.J., & Ashill, N. (2006). The determinants of students' perceived learning outcomes and satisfaction in university online education: an empirical investigation. *Decision Sciences Journal of Innovative Education, 4*(2), 215–235.

Geisinger, K. F. (2016). 21st century skills: What are they and how do we assess them? *Applied Measurement in Education, 29*(4), 245–249.

Gipps C. V. (2005). What is the role for ICT-based assessment in universities? *Studies in Higher Education*, *30*, 171–180.

González-Gómez, F., Guardiola, J., Martín-Rodríguez, O. & Montero-Alonso, M.A. (2012), Gender differences in e-learning satisfaction. *Computers & Education*, *58*(1), 283–290.

Gray, D., Ryan, M., & Coulon, A. (2004). The training of teachers and trainers: innovative practices, skills and competencies in the use of e-learning. *European Journal of Open, Distance and E-Learning*, *2004*(2) 11.

Greenstein, L. (2012). *Assessing 21st century skills: A guide to evaluating mastery and authentic learning*. Corwin, a SAGE Company.

Hertel, G., Naumann, S., Konradt, U., & Batinic, B. (2002). Personality assessment via internet: Comparing online and paper-and-pencil questionnaires. In B. Batinic & U. D. Reips & M. Bosnjak (Eds.), *Online Social Science*, (pp. 115–133).

Ho, C.L., & Dzeng, R.J. (2010). Construction safety training via e-learning: learning effectiveness and user satisfaction. *Computers & Education*, *55*(2), 858–867.

Hoferichter, F., Raufelder, D., Ringeisen, T., Rohrmann, S., & Bukowski, W.M. (2015), Assessing the multi-faceted nature of test anxiety among secondary school students: An English version of the German test anxiety questionnaire: PAF-E, *The Journal of Psychology*, 1–23. http://dx.doi.org/10.1080/00223980.2015.1087374

Hoskins, S. L., & Van Hooff, J. C. (2005). Motivation and ability: Which students use online learning and what influence does it have on their achievement? *British Journal of Educational Technology*, *36*(2), 177–192.

Jeong H. (2012). A comparative study of scores on computer-based tests and paper-based tests. *Behaviour & Information Technology*, *33*, 410–422.

Kelly, H.F., Ponton, M.K., & Rovai, A.P. (2007). A comparison of student evaluations of teaching between online and face-to-face courses. *The Internet and Higher Education*, *10*(2), 89–101.

Lemos, S., & Nueza, P. (2012), Students expectation and satisfaction in postgraduate on-line courses, ICICTE 2012 Proceedings, 568–580.

Luecht, R. M. (2013). An introduction to assessment engineering for automatic item generation. In M. J. Gierl & T. M. Haladyna (Eds.), *Automatic item generation: Theory and practice* (59–78). Routledge.

Martín-Rodríguez, Ó., Fernández-Molina, J.C., Montero-Alonso, M.Á., & González-Gómez, F. (2015), The main components of satisfaction with e-learning. *Technology, Pedagogy and Education*, *24*(2), 267–277.

McFadden A. C., Marsh G. E., & Price B. J. (2001). Computer Testing in Education. *Computers in the Schools*, *18*, 43–60.

Mills, C. N., & Stocking, M. L. (1996). Practical issues in large-scale computerized adaptive testing. *Applied Measurement in Education*, *9*, 287–304.

Mills C. N., Potenza M. T., Fremer J. J., & Ward W. C. (2005). *Computer-based testing: Building the foundation for future assessments*, Materials evaluation and design for language teaching. Edinburgh, UK: Edinburgh.

Miri, C. (November, 2016). Personal communication. DOI: 10.13140/RG.2.2.32040.88326. https://www.researchgate.net/publication/324208499.

Neemati, N., Hooshangi, R., & Shurideh, A. (2014), An investigation into the learners' attitudes towards factors affecting their exam performance: A case from Razi University, *Procedia-Social and Behavioral Sciences*, *98*, (May), 1331–1339.

Nikou S., & Economides A. A. (2013). Student achievement in paper, computer/web and mobile based assessment, *BCI (Local)*, 55, Part B, 1241–1248, https://doi.org/10.1016/j.chb.2015.09.025

Nikou, S.A., & Economides, A.A. (2016), The impact of paper-based, computer-based and mobile-based self-assessment on students' science motivation and achievement, *Computers in Human Behavior*, *55*, February, 1241–1248.

Paechter, M., & Maier, B. (2010), Online or face-to-face? Students' experiences and preferences in e-learning. *The Internet and Higher Education, 13* 4, 292–297.
Pellegrino, J., Chudowsky, N., & Glaser, R. (2001). *Knowing what students know: The science and design of educational assessment.* National Academy Press.
Piaw, C. Y. (2012). Replacing paper-based testing with computer-based testing in assessment: Are we doing wrong? *Procedia-Social and Behavioral Sciences*, 64, 655–664.
Piaw Chua, Y. (2012). Effects of computer-based testing on test performance and testing motivation. *Computers in Human Behavior, 28*(5), 1580–1586.
Quellmalz, E. (2014). Computer-Based Assessment. In: Gunston, R., Ed., *Encyclopedia of Science Education SE-44-2*, Springer, 1–6.
Raymond, C., & Usherwood, S. (2013). Assessment in simulations. *Journal of Political Science Education, 9*(2), 157–167.
Rehmani, A. (2003). Impact of Public Examination System on Teaching and Learning in Pakistan. Retrieved November 10, 2016. 'ANTRIEP': New Delhi, January June 2003 Vol. 8, No 1.URL (http://www.antriep.net/html/Antriep%20jan-june%202003.pdf).
Robles, M., & Braathen, S. (2002). Online assessment techniques. *Delta Pi Epsilon Journal, 44*(1), 39–49.
Rollings-Carter F. (2010, May 28). *Performance assessments versus traditional assessments.* Available: http://www.learnnc.org/lp/editions/linguafolio/6305 [
Russell M., & Haney W. (2000). Bridging the gap between testing and technology in schools, *Education Policy Analysis Archives, 8*, 19.
Sands, W. A., Waters, B. K., & McBride, J. R. (Eds.). (1997). *Computerized adaptive testing: From inquiry to operation.* American Psychological Association.
Segall N., Doolen T. L., and Porter J. D., (2005). A usability comparison of PDA-based quizzes and paper-and-pencil quizzes. *Computers & Education, 45*, 417–432.
Sun, P., Tsai, R., Finger, G., Chen, Y., & Yeh, D. (2008). What drives a successful e-Learning? An empirical investigation of the critical factors influencing learner satisfaction. *Computers & Education, 50*(4), 1183–1202.
Swan, K. (2001), Virtual interaction: design factors affecting student satisfaction and perceived learning in asynchronous online courses *Distance Education, 22*(2), 306–331.
Terzis V., & Economides A. A. (2011) Computer based assessment: Gender differences in perceptions and acceptance. *Computers in Human Behavior, 27*, 2108–2122.
Thelwall, M. (2000). Computer-based assessment: a versatile educational tool. *Computers and Education, 34*(1), 37–49
Thompson, D. (2008). Integrating graduate attributes with assessment criteria in business education: Using an online assessment system. *Journal of University Teaching and Learning Practice 5*(1), 4.
Triantafillou E., Georgiadou E., & Economides A. A. (2008). The design and evaluation of a computerized adaptive test on mobile devices. *Computers & Education, 50*, 1319–1330.
Volman, M. (2005). A variety of roles for a new type of teacher educational technology and the teaching profession. *Teaching and Teacher Education, 21*(1), 15–31.
Vonderwell, S., & Boboc, M. (2013). Promoting Formative Assessment in Online Teaching and Learning. *TechTrends: Linking Research & Practice to Improve Learning, 57*(4), 22–27.
Voogt, J., & Roblin, N. P. (2012). A comparative analysis of international frameworks for 21st century competences: Implications for national curriculum policies. *Journal of Curriculum Studies, 44*(3), 299–321.
Zara, A. R. (1994). An overview of the NCLEX/CAT beta test. Paper presented at the meeting of the American Educational Research Association, New Orleans.

6 Integration of Digital Technologies for Constructing Influential Learning during the COVID-19 Pandemic in Pakistan

Muhammad Mujtaba Asad and Kanwal Aftab
Sukkur IBA University, Pakistan

Fahad Sherwani
National University of Computer and Emerging Sciences, Pakistan

Al-Karim Datoo and Zahid Husaain Khand
Sukkur IBA University, Pakistan

CONTENTS

DOI: 10.1201/9781003102298-6

TABLE 6.1
Research Contributions in ICTs in Education

Author	Year	Contribution
Dede	2000	This article was written on the Developing impacts of information technology on school curriculum.
Kozma	2003	This article is based on the Global perspective of technology, innovation, and educational change.
Tinio	2013	ICT in Education.
Tubin	2006	This article was written on Different varieties of ICT presentation and technology applications

6.1 INTRODUCTION

ICT refers to the information and communication technologies which are well defined as a set of various technical gears and capitals used to transmit and produce, distribute, supply, and achieve information of this introduction (Tinio, 2003). ICT is seen as a tool to help schools meet the age change from industry to information (Kozma, Kozma, & McGhee, 2003). However, the efficiency, complexity, and strength of ICT turn this change into an obstacle (Tubin, 2006). ICT is defined as an established set of gears to offer, support, and strengthen education reforms based on the educational needs of the information-based society (Dede, 2000) (Table 6.1).

6.2 ICT IS A MEANS OF CONSTRUCTING INFLUENTIAL LEARNING

ICT can donate to generating an influential knowledge situation in many behaviors. ICT provides access to large amounts of information by using a large number of information sources and displaying information from many angles, thereby increasing the reality of the learning environment. ICT can also promote understanding of difficult procedures through models that pay to exclusive education surroundings (Smeets, 2005). ICT can encourage collaborative learning and thinking about content (Susman, 1998). In addition, ICT can be used as a tool for differentiated courses, providing opportunities to adapt learning content and tasks to each student's needs and abilities and providing specific feedback (Mooij, 1999). ICT enhances students' creativity and collaboration, practitioners report that they rarely or never do, is still high (Davies & Pittard, 2009). ICT education includes an outline modification that enables new visions and new types of understanding. Such standard shifts require geography teachers who can handle technical changes in the classroom. Therefore, there is a need to move in the direction of ICT literacy by including ICT into school constructions and learning organizations (Nato & Benjamin, n.d.). In the modern world, the use of ICT has helped humans to develop many things and improve their thoughtful statement, and difficult resolving assistances concluded an extensive variety of software and input strategies (Thomas & Stratton, 2006) (Table 6.2).

TABLE 6.2
ICTs for Constructing Influential Learning

Author	Year	Contribution
Davies & Pittard	2007	Connecting technology. Basically, it is a review. It discusses the role of technology in education and skills.
Mooij	1998	Gives us a clear guideline to Educational use of ICT in Teaching.
Nato & Benjamin	2013	Focuses on the natures of ICT Materials which are obtainable for Schooling of Geography in Secondary.
Smeets	2005	Impact of ICT and how it can give dominant education settings in primary education
Susman	1998	Supportive learning: A review of features that raise the usefulness of supportive computer-based teaching.
Thomas & Stratton	2006	ICT in physical education: A national audit of apparatus, use, teacher approaches, provision, and training.

6.3 INFLUENCE OF ICT INCORPORATION IN EDUCATION

The influence of ICT on an information-based society has changed dramatically. In terms of form and content, the goal of knowledge to enter a wider society was tremendous and a multiplier; education was one of the broader influences and developments that took place there (Hernandez, 2017). Institute is one of the places where equipment has the highest impact. This is where everyday life becomes part of the school, which has an impact on the role of the teacher. ICT integration into education has developed a procedure, and the results go far beyond the technical tools that provide nutrition for educational environments (Para Mosquera, 2012). The changes in ICT have made informative apparatuses that can advance the excellence of learners' education and totally modify the technique through which information is learned, achieved, and understood (Aguilar, 2012). As portion of the part engage in recreation by each educational intermediary institution, learners are now using technical apparatuses to promote education. This growth instigated the arrival of advanced machine, televisions, tape recorder, and other devices. On the other hand, progress has been made in that scientific assets have become informative incomes. In this case, work to improve the level of learning needs to incorporate technology into education. The usage of ICT means separation from old-style media, wooden panels, pencils, etc., plus it plays a teaching role on the basis of updating the knowledge of teaching methods according to people's educational needs and current needs (Granados, 2015). Meanwhile, teaching is an imperative characteristic of human life; digitalization has become a revolution, so it is combined with ICT to make a fresh cultural background that allows students to be accountable for their own knowledge. Period and flexibility perform an important character with the emergence of new technologies, emerging education, and teaching paradigms (Suarez & Najar, 2014) (Table 6.3).

TABLE 6.3
Influence of ICT Incorporation in Education

Author	Year	Contribution
Aguilar	2012	Knowledge and Facts and Statement Technologies
Granados	2015	ICT in the instruction of arithmetical approaches
Hernandez	2017	This article analyzed the Influence of ICT on Teaching and what are its Challenges and Perspectives.
Mosquera	2012	ICT, facts, teaching, and technical assistances in teacher preparation
Suarez & Najar	2014	Development of evidence and statement skills in the teaching–learning procedure

6.4 ADAPTABILITY OF ICTS AMONG TEACHERS

Teachers have an actual significant character to show in society. The character they composed in the education development is essential to elementary teaching, especially in Third Creation republics. One of the main problems faced by the schooling structure is the lack of qualified teachers who are well informed or skillful in the custom of information technology (Singh & Chan, 2014). Research on primary school teachers' "perceptions of ICT and teaching issues has shown that teachers" ICT ideas are shaped by their involvement in a wide range of cultural and social areas affecting their professional fields and environments. Loveless (2003) collected principles about teacher ICT in three groups: the first one is ICT in civilization: Teachers spoke about the "information society" and its influence on children's future education and lives. The second one is ICT Qualifications: Teachers communicated about ICT skills or "literacy information" that children necessity as a focus and as an interdisciplinary device. And the last one is ICT in schools: Teachers spoke about how deficiency of technology and shortage of properties in schools affect incorporation. These ideas replicate the constant debate over the significance of ICT in teachers' work, and seeing them as a source of tension rather than concern is more practical for the ongoing denotation (Loveless, 2003). Student–teacher learning principles were described as the starting point for responding to children's ideas, compromising, and compromising, answering to youngsters' ideas, and "not giving too much supervision because of drowning" (Loveless, Burton & Turvey, 2006) (Table 6.4).

TABLE 6.4
Adaptability of ICTs Among Teachers

Author	Year	Contribution
Loveless	2003	Communication between primary teachers' insights of ICT and their schooling
Loveless, Burton	2006	In this article authors develop theoretical backgrounds for creativeness, ICT, and educator teaching.
Singh & Chan	2014	Educator willingness on ICT addition in teaching–learning

6.5 CLASSIFICATION OF ICTS IN EDUCATION

ICTs in education are classified into three main groups, E-Learning, Blended Learning, and Distance Education.

6.5.1 E-Learning

Electronic learning or e-learning is a general term for computer-assisted learning. It is fundamentally related to the field of Advanced Learning Technology (ALT), which compacts both technologies and related methods of learning using network and/or multimedia technologies (Kumar, 2008). E-learning is primarily defined as the use of computer network technology over the Internet or on the Internet to provide information and education to individuals through computer-based learning, online learning, distributed learning, or sometimes used web-based learning, e-learning will become more and more prevalent as a preferred term in organizations (Welsh, Wanberg, Brown & Simmering, 2003). Although e-learning is most often related to advanced teaching and institutional education, e-learning covers all levels and informal learning, whether it is full or partial usage of the information network Internet (LAN) or (WAN) (Tinio, 2003).

6.5.2 Blended Learning

Blended learning is an arrangement of numerous learning methods. It is typically used to define a condition in which unlike supply procedures are mutual to offer a specific course. These approaches can contain face-to-face classrooms, a mix of self-study and online classrooms (Kumar, 2008). These state to a learning classic that association's old-style classroom exercise with e-learning solutions. For example, students in outdated classes can be given paper and online resources, online discussion meetings can be controlled from end-to-end chats with teachers, and class mailing lists can be subscribed (Tinio, 2003). Blended learning is both simple and complex. The simplest form of blended learning is a thoughtful fusion of the face-to-face learning experience and the online learning experience. The concept of integrating the advantages of simultaneous (face-to-face) and asynchronous (text-based Internet) learning activities is largely intuitively attractive (Garrison & Kanuka, 2004).

6.5.3 Distance Learning

Distance learning is a way to provide learning opportunities, and it is characterized by "the wealth of common learning", that is, it is recognized by an institution or organization in some way; it uses various media, including printing and electronics; Interaction; possibility of face-to-face meeting from time to time; and division of labor specializing in course production and presentation (Tinio, 2003). This is a kind of education where students can effort alone at family or at workplace and links per faculty members also other students through message, electronic forums, cinematographic conferences, discussion rooms, immediate messaging, and other

TABLE 6.5
Classification of ICTs in Education

Author	Year	Contribution
Garrison & Kanuka	2004	Discovery of transformative prospective of blended learning in upper schooling.
Kumar	2008	Merging of ICT and Teaching.
Tinio	2003	ICT in Schooling.
Welsh, Wanberg	2003	E-learning: developing usages, experiential consequences, and upcoming instructions.

computer-based communication methods. This is also called open learning. Greatest distance education programs comprise computer-based education (CBT) arrangements and announcement apparatuses for producing computer-generated classrooms. Since the Internet and World Wide Web can be retrieved from nearly any PC stand, they form the foundation of many distance education organizations (Kumar, 2008) (Table 6.5).

6.6 GLOBAL TRANSMISSION OF COVID-19

For the first time since December 2019, pneumonia of unknown etiology has spread in Wuhan, Hubei Province, China. Following the outbreak, a novel corona virus, SARS-CoV-2, was identified as the contributing virus for epidemic disease in China and other parts of the world, according to the World Health Organization (WHO, 2020). Throughout the COVID-19 outbreak, it is overbearing to understand in what way the inhabitants in extremely exaggerated countries such as China survived with such a major tragedy (Zhang & Ma, 2020).

- *Symptoms*
  Symptoms can contain temperature, cough, and littleness of breath. In more thoughtful cases, the contagion can reason pneumonia or living problems. Additional hardly, the illness can be mortal. These symptoms are related to (flu) or chills, which are considerably more common than COVID-19. Therefore, analysis is compulsory to indorse whether someone is COVID-19 (WHO, 2020).
- *Death rate*
  On 20 March 2020, 166 countries and 274,180 patients were confirmed to be infected, 11,375 died and 87,991 recovered. The mortality rate was considered between 0.39 and 4%, but this is subject to the age of the infected person enduring and is abundant in those over 70 years old. The inhabitants most possible to need mechanical exposure to air are the elderly and those with associated comorbidities (especially cardiovascular disease and hypertension, followed by diabetes mellitus) with an estimated mortality of about 15–49% (Kowalski et al., 2020).

TABLE 6.6
Global Transmission of COVID-19

Author	Year	Contribution
Kowalski, Sanabria	2020	Head and neck surgery practice during COVID-19 pandemic.
UNISEF	2020	Contagion preclusion and controller throughout health care when COVID-19 is supposed.
Viner, Russell	2020	Institute closure applies during coronavirus outbreaks including COVID-19
WHO	2020	Transformation of COVID-19, its symptoms and Death rates
Zhang & Ma	2020	COVID-19 pandemic on mental health and quality of life among local residents.

- *Preventive actions*
  Other respiratory infections, such as the flu or the common cold, are important to reduce the spread of public health diseases. Public health measures are daily safety measures that include:
  ✓ Remaining at homespun when you are in bad taste
  ✓ Wrapping your mouth and nose with bent elbow or tissue when coughing or sneezing
  ✓ Disposing of used tissue instantly
  ✓ Washing hands often with soap and water
  ✓ Washing affected surfaces and objects regularly
  ✓ Social distancing

As we learn more about COVID-19, public health officials may recommend additional measures (UNISEF, 2020) to reduce COVID-19 shipments. But the national shutdown began. These measures appear to be largely based on the assumption that the benefits of influenza outbreaks are also valid for COVID-19 (Viner et al. 2020) (Table 6.6).

6.7 COVID-19 IN PAKISTAN

In March 2020, Pakistan had 1179 cases of COVID-19, with most 421 cases from Sindh, 394 cases, 131 cases, 123 cases, 84 cases, 25 cases, and 01 cases from Punjab, Balochistan, Khyber Pakhtunkhwa, Gilgit-Baltistan, Islamabad Capital Territory, and Azad Jammu and Kashmir, respectively (Raza, Rasheed & Rashid, 2020). Like the rest of the world, Pakistan's public health sector aims to strengthen national structures to curb the spread of global diseases; regulators understand that biological risks will not only have an impact on environmental health but will also cause far-reaching social and economic damage (Hyder, 2020). With the unprecedented popularity of COVID-19, education has been particularly hit. The Pakistani government temporarily closed educational institutions to curb the spread of the epidemic. Millions of learners are not in schools, colleges, or universities. Due to this major change, the dropout rate is expected to increase because it will force many low-cost schools to close permanently. As a result, many students are admitted to public education institutions, which are already overcrowded and lack the basic facilities to provide services to students who are already there (Hyder, 2020). On March 23, Pakistan announced a two-week blockade of the country, suspended all domestic transportation, ceased international air travel, and called for troops to handle the situations (Khan, Niazi, & Saif, 2020). After the blockade, all institutes were closed until additional notice, and board examinations (Basic, Intermediate, and GCSE) were also suspended. Selected English secondary schools are arranging for computer-generated schools, some of which are still distributing Chrome books to students, while public and low-income private schools have no strategies for computer-generated home schools (Hyder, 2020). On 26 March, Federal Minister for Federal Education and Professional Training, Shafqat Mehmood announced that universities and other educational institutions would be closed until 31 May (Khan, et al. 2020). Like the rest of the world, the COVID-19 pandemic has had a negative impact on the education of millions of students in Pakistan. Parents and teachers are increasingly worried: what impact may the long-term closure of educational institutions have on the short-term learning and long-term success of students? There is no doubt that the impact of school, college, and university suspensions on student learning is negative and disproportionate. Compared with students from wealthy backgrounds and urban areas, the suspension of classes has a greater negative impact on the education of learners from low-income families or rural parts (Hazir-ullah a, 2020). Suspension of classes for low-income families means that children should help their families. The children worked with their families in agriculture, firewood, and cattle breeding. Some are even forced to engage in child labor. Similarly, there is no educational environment at home; children and parents have insufficient interest in school; family poverty; the pressure of family responsibilities; and the absence of fathers due to working outside the home will cause these children to make up for the loss of learning. If the schools are closed for more than May 31, it will lead to excessive loss of learning (Hazir-ullah b, 2020) (Table 6.7).

TABLE 6.7
COVID-19 in Pakistan

Author	Year	Contribution
Hazir-ullah a,	2020	COVID-19 and education.
Hazir-ullah b,	2020	COVID-19 and Education.
Hyder	2020	SHORT NOTES ON THE ECONOMY DURING THE COVID-19 CRISIS.
Khan, Niazi	2020	Universities unprepared for switching to remote learning.
Raza, Rasheed	2020	Broadcast Possible and Sternness of COVID-19 in Pakistan.

6.8 TEACHING AND LEARNING EXPERIENCES USING ICT IN EDUCATION DURING COVID-19 IN PAKISTAN

With the popularity of COVID-19, many college campuses have previously offered in-person courses online (Sutton, 2020). The Higher Education Commission (2020) has initiated public sector universities to make online lectures to moderate any intervention in the provision of education. Scholars are predictable to be gifted to access the Internet connection and computer so that they can retrieve information. Even in the near future, low-income public / private schools can mobilize their teachers to introduce online lectures (Hyder, 2020). For many such elite schools and universities, this is a minor jump forward. Their students educated their parents to use laptops / desktops with Internet access at home. In fact, this is not a compensation for the rich and meaningful teaching and learning experience in the classroom (Hazir-ullah a, 2020). However, younger elementary and middle school students need parental supervision and guidance when trying to complete tasks assigned by teachers in a virtual school; but many of them are first-generation students, so they may not be able to get guidance at home: these students may need additional help, and the state government may be able to provide these help through public service education broadcasts on radio and television (Hyder, 2020). However, for some reasons, online courses make up for the loss of learning to some extent, because these online courses have always made students interested in content. In addition to the development of content, they also provide knowledge on what way to custom technology for education. They communicate with educators and gather courses / homework through emails, websites, and engage in video conferences. Students use the same technology, educational software, and applications to form learning groups (Hazir-ullah a, 2020). During the COVID-19 crisis, the sudden shift to digital learning posed some challenges to the system because most students did not have their own computers or Internet facilities (Khan et al., 2020). The increase in Internet usage has also triggered warnings about privacy leaks, which may increase in the coming days. Since the service model provided by the third party does not comply with the existing licensing system, privacy issues have always been a challenge in Pakistan. "At present, the" Electronic Crime Prevention Act "is promoting the authorities to protect Internet users in Pakistan. But in emergency situations, standard procedures do not work (Butt, 2020). Technical topics and laboratory time cannot be taught on zoom or other video streaming software. It is worth noting that many students were in trouble due to the blockade. After the blockade, they were unable to go home or stay in hotels in other cities or universities. In this circumstance, they cannot probably to join compulsory online courses (Noor-ul-ain, 2020). They are also losing social interaction. They are losing sports and extracurricular activities as important as the course content (Hazir-ullah a, 2020). Leaving aside the lack of social communication, deprived Internet construction, sound, absence of self-governing knowledge assistances, or emotional maturity for children and young children to uphold care, it can be concluded that students in these urban elite schools have more disadvantaged students than SES who participate in public education Learning advantages General educational institutions, especially those in rural areas (Hazir-ullah b, 2020). At present, Pakistani universities face many challenges in arranging online courses because their

TABLE 6.8
Teaching and Learning Experiences in COVID-19

Author	Year	Contribution
Butt	2020	Pakistan's Internet usage flows amongst COVID-19 lockdown
Hazir-ullah a,	2020	COVID-19 and teaching:
Hazir-ullah b,	2020	COVID-19 and Schooling
HEC	2020	Asks universities to start online teaching. *The News International.*
Hyder	2020	SHORT NOTES ON THE ECONOMY DURING THE COVID-19 CRISIS.
Khalid	2020	COVID-19 and educational crises
Khan, Niazi	2020	Universities unprepared for switch to remote learning.
Manzoor	2020	Online education and encounters of COVID-19 for enclosure of PWDs in advanced teaching.
Noor-ul-ain	2020	Students disappointed with online teaching system amid COVID-19
Sutton	2020	Offer faculty a convenient attendant to meeting online course convenience supplies.

management is not prepared in advance. Serious problems, such as lack of teacher preparation, as well as the lack of necessary funds and resources, have become an obvious obstacle (Khalid, 2020). Online courses will bring more challenges for people with disabilities to successfully integrate into higher education institutions. The main challenge of online teaching will be assessment and evaluation; it will be a complicated process for online teachers to arrange individual assignments for students with disabilities according to the degree and type of disability. In addition, honesty and integrity in providing online feedback will also become a challenge for teacher evaluation (Manzoor, 2020) (Table 6.8).

6.9 CONCLUSION

The above discussion from literature concludes that transformation of learning environment through ICTs during COVID-19 in Pakistan has negative impact on student's learning. Both students and teachers face many challenges, such as loss of social interaction ability, poor Internet connection, noise, lack of independent learning skills for children and toddlers, and their management is not prepared in advance. Serious problems, such as lack of teacher preparation, as well as the lack of necessary funds and resources, have become an obvious obstacle. Many people believe that ICT is a catalyst for change, changing working conditions, access to information processing and communication, teaching methods, learning methods, scientific research, and access to information technology, and it can help teachers present his / her teaching in an attractive way, but online teaching also brings challenges to assessment and evaluation; it will be a complicated process for online teachers to arrange personal assignments for students. In addition, honesty and integrity in providing online feedback will also become a challenge for teacher evaluation in the online teaching process.

Finally, the developed countries have used this technology appropriately for education in remote localities. The authorities must consider these challenges in order to make successful and fair inclusion. For students who cannot attend online courses for some reason, they must also be flexible. In addition, universities must ensure mental health by providing health and safety guidelines and responding to pressures caused by social alienation. Moreover, the University Counseling Center can expand the range of services online to facilitate students when needed. Other offices, such as exam and admission offices, can also be upgraded online to facilitate personnel.

REFERENCES

Aguilar, M. 2012. Learning and information and communication technologies: Towards new educational scenarios. *Latin American Journal of Social Sciences, Children and Youth, 10* (2), 801–811.

Butt. (2020, April 16). *Pakistan's internet use surges amid COVID-19 lockdown* https://www.aa.com.tr/en/asia-pacific/pakistan-s-internet-use-surges-amid-covid-19-lockdown/1807118

Davies, S., & Pittard, V. (2009). Harnessing technology review 2009. The role of technology in education and skills.

Dede, C. (2000). Emerging influences of information technology on school curriculum. *Journal of Curriculum Studies*, *32*(2), 281–303.

Garrison, D.R., & Kanuka, H. (2004). Blended learning: Uncovering its transformative potential in higher education. *The Internet and Higher Education*, *7*(2), 95–105.

Granados, A. (2015). ICT in the teaching of numerical methods. *Sophia Education*, *11*(2), 143–154.

Hazir-ullah a. (2020, May 1). *COVID-19 and education: Unequal learning loss*. The Nation. https://nation.com.pk/01-May-2020/covid-19-and-education-unequal-learning-loss

Hazir-ullah b. (2020, May 7). *COVID-19 and Education: Unequal learning loss. Pakistan today*. https://www.pakistantoday.com.pk/2020/05/07/covid-19-education-unequal-learning-loss/

Hernandez, R.M. (2017). Impact of ICT on education: Challenges and perspectives. *Journal of Educational Psychology-Propositos y Representaciones*, *5*(1), 337–347.

Hyder, A. (2020). *Short notes on the economy during the covid-19 crisis*. SSRN.

Khalid. (2020, April 30). *COVID-19 and educational crises* https://www.pakistantoday.com.pk/2020/04/30/covid-19-educational-crises/

Khan, A. A., Niazi, S., & Saif, S. K. (2020, 26 March). *Universities unprepared for switch to remote learning*. https://www.universityworldnews.com/post.php?story=20200326141547229

Kowalski, L.P., Sanabria, A., Ridge, J.A., Ng, W.T., de Bree, R., Rinaldo, A., Takes, R.P., Mäkitie, A.A., Carvalho, A.L., Bradford, C.R., & Paleri, V. (2020). COVID-19 pandemic: Effects and evidence-based recommendations for otolaryngology and head and neck surgery practice. *Head & Neck*, *42*, 1259–1267.

Kozma, R., Kozma, R.E., & McGhee, R. (2003). *Technology, innovation, and educational change. A global perspective*. Academia. https://d1wqtxts1xzle7.cloudfront.net/62234760/kozma_ll_article-with-cover-page.pdf?Expires=1623047296&Signature=Wn~LRmazGoCOPII-pkivRoHPQc9CqL9a0rzTZ3SdXQbj9AE-TSlWp-gELqChqZ0sL6bqLMsEGI6-62aN6K9FGmkWPRXwctpTfQPFe57ah0GMslapSskJp~VWRB9JSyv4w4-LK1XAMqbOy6h64EVu7b9yrQSeimyP26OfzU~nhauEh8BBQSmJABuVW5sFiZ7eABnkfGFormcsM30OqRm4GuR5NVXUCLj7jt-0z19cRn6AMIJzJN04Pgl4HpqK0cFcgy4CxDNnNIMLJEqeZerr3jn3GJN4sQG~nc8LMaCETD2NuZKaM8ZMRmpch8Qe9T4jb~TAUL5-F3WMVhEzw5uX5Cg__&Key-Pair-Id=APKAJLOHF5GGSLRBV4ZA

Kumar, R. (2008). Convergence of ICT and Education. *World Academy of Science, Engineering and Technology*, *40*(2008), 556–559.

Loveless, A.M. (2003). The interaction between primary teachers' perceptions of ICT and their pedagogy. *Education and Information Technologies*, *8*(4), 313–326.

Loveless, A., Burton, J., & Turvey, K., (2006). Developing conceptual frameworks for creativity, ICT and teacher education. *Thinking Skills and Creativity*, *1*(1), 3–13.

Manzoor. (2020, April 15). *Online teaching and challenges of COVID-19 for inclusion of PWDs in higher education.* https://dailytimes.com.pk/595888/online-teaching-and-challenges-of-covid-19-for-inclusion-of-pwds-in-higher-education/

Mooij, T. (1999). Guidelines to Pedagogical Use of ICT in Education. Paper presented on the eighth conference of the 'European Association for Research on Learning and Instruction' (EARLI). Göteborg, Sweden.

Nato, L.W., & Benjamin, O. (2013). Types of ICT Materials available for Teaching of Geography in Secondary Schools in Rongo District.

Noor-ul-Ain Ali. (2020. April 2). *Students disappointed with online teaching system amid COVID-19.* https://dailytimes.com.pk/587446/students-disappointed-with-online-teaching-system-amid-covid-19/

Parra Mosquera, C. A. (2012). ICT, knowledge, education and technological skills in teacher training. *Nomads*, (36), 145–159.

Raza, S., Rasheed, M.A., & Rashid, M.K., (2020). *Transmission potential and severity of covid-19 in Pakistan.* Preprints https://www.preprints.org/manuscript/202004.0004/v1

Singh, T.K.R., & Chan, S. (2014). Teacher readiness on ICT integration in teaching-learning: A Malaysian case study. *International Journal of Asian Social Science*, *4*(7), 874–885.

Smeets, E. (2005). Does ICT contribute to powerful learning environments in primary education?. *Computers & Education*, *44*(3), 343–355.

Suarez, N.E.S., & Najar, J.C. (2014). Evolution of information and communication technologies in the teaching-learning process. *Vínculos Magazine*, *11*(1), 209–220.

Susman, E.B. (1998). Cooperative learning: A review of factors that increase the effectiveness of cooperative computer-based instruction. *Journal of Educational Computing Research*, *18*(4), 303–322.

Sutton, H. (2020). Offer faculty a handy guide to meeting online course accessibility requirements. *Disability Compliance for Higher Education*, *25*(10), 9–9.

Thomas, A., & Stratton, G. (2006). What we are really doing with ICT in physical education: a national audit of equipment, use, teacher attitudes, support, and training. *British Journal of Educational Technology*, *37*(4), 617–632.

Tinio, V.L. (2003). *ICT in Education.* Tbilisi State University.

Tubin, D. (2006). Typology of ICT implementation and technology applications. *Computers in the Schools*, *23*(1–2), 85–98.

Viner, R.M., Russell, S.J., Croker, H., Packer, J., Ward, J., Stansfield, C., Mytton, O., Bonell, C., & Booy, R. 2020. School closure and management practices during coronavirus outbreaks including COVID-19: A rapid systematic review. *The Lancet Child & Adolescent Health*, *4*, 397–404

Welsh, E.T., Wanberg, C.R., Brown, K.G., & Simmering, M.J. (2003). E-learning: Emerging uses, empirical results and future directions. *International Journal of Training and Development*, *7*(4), 245–258.

World Health Organization. (2020). *Infection prevention and control during health care when COVID-19 is suspected.* https://www.who.int/emergencies/diseases/novelcoronavirus2019/technical-guidance/infection-prevention-and-control.

Zhang, Y., & Ma, Z.F. (2020). Impact of the COVID-19 pandemic on mental health and quality of life among local residents in Liaoning Province, China: A cross-sectional study. *International Journal of Environmental Research and Public Health*, *17*(7), 2381.

7 Applications of ICT
Pathway to Outcome-Based Education in Engineering and Technology Curriculum

Prashant Gupta, Trishul Kulkarni
Maharashtra Institute of Technology, India

Vishal Barot
LDRP Institute of Technology and Research, India

Bhagwan Toksha
Maharashtra Institute of Technology, India

CONTENTS

DOI: 10.1201/9781003102298-7

7.1 INTRODUCTION

The term outcome-based education (OBE) was first coined by William G. Spady in 1988 and stimulated the simultaneous creation of Washington Accords in order to support globalization for the joint appreciation of engineering qualifications (Spady, 1994). Harden further reinforced OBE as a performance-oriented approach at the groundbreaking mode of curriculum development (Harden, 1999). The Washington Accord has 21 full signatories and 7 provisional signatories as of date (International Engineering Alliance, 2020). It necessarily directs to a student-centric teaching–learning methodology that is objective, based on terms of content creation, delivery and assessment. Also, the focus in OBE at different levels is on the assessment of student performance or outcomes which has become the need of the hour with an increased expectancy of effectiveness in the working environments. OBE is already practiced in various fields of education, with medicine as one of the skill-oriented, professional educational sectors. Due to the introduction of OBE in teaching–learning activities, the medical sector has reaped good results by being more student-centric in the past two decades (Haque, 2017; Paterson Davenport et al., 2005).

The use of Information and Communication Technology (ICT) is gradually catching up in all spheres of human life largely due to the swift nature of job completion on offer with lesser human efforts. The implementation of both ICT and OBE in an engineering education setup has challenges of its own. Some of these challenges include an investment of the stakeholders, resources available, etc. for ICT and know-how involved along with human efforts involved in OBE. Out of all the alternatives that may be able to help in the effective implementation of OBE, ICT is capable of assisting the current teaching–learning setup with a faster rate of deployment of currently practiced activities done in a conventional way without the use of technology.

ICT tools that supplement engineering and technology education comprise of Learning Management Systems (LMSs), which are increasingly being used across the globe. They are cloud based and are either paid or open-source free, e.g. Moodle, Blackboard, Google Classroom, etc. There are free versions available, but paid versions are largely fluent, user friendly and free of bugs. Furthermore, the inefficiency of conventional engineering teaching–learning setup, wherein the current examination framework is unable to cope up with increasing demands of higher education such as OBE, requires reforms in the assessment which can be addressed with the use of ICT. It also has a capability to tap into higher cognitive levels of learning such as Analyze, Evaluate and Create as the need of the hour is learning to have a command over these higher levels of cognitive knowledge. The learners upon completing their education will be expected to perform tasks/activities relevant to these cognitive levels on an everyday basis as a part of their work since their employer will constantly look to keep itself relevant with the advancements in the competitive market. Furthermore, it is important that these levels are well practiced while in work, which in turn inculcates the entrepreneurship attitude/skills in the engineer. However, the challenges faced by teachers for use of current ICT arrangements is one of the reasons out of many, such as limited resource availability largely at the learners end along with high-speed internet connectivity, responsible for limiting its widespread use. The instructors face challenges right from curriculum design, content creation/development and its delivery with limited knowledge of resources and the time required to be invested in doing the activities. The traditionally trained instructors need proper training which is seldom given as they are expected to learn by themselves in today's web-based competitive world.

The role of ICT in engineering and technology education, its use in augmenting OBE, LMS and its use in implementing complementary models such as Analysis, Design, Development, Implementation, and Evaluation (ADDIE), Technological Pedagogical Content Knowledge (TPACK) and addressing higher cognitive levels is highlighted and the significance of human–computer interaction in easing the teaching–learning process along with the use of artificial intelligence (AI), virtual environments, augmented reality, etc. is discussed and elaborated in this chapter.

7.2 ROLE OF ICT IN ENGINEERING EDUCATION

ICT refers to managing and broadcasting media, telecommunications, audio–visual transmission, processing technology and network-oriented control enabled with an overview of functionalities. It is an umbrella encompassing any device/application capable of communicating of which an exhaustive list includes TV, mobile phones, radio, computer systems, network software and hardware, satellite systems and assisting services/applications of the same, e.g. distance learning with the help of video/teleconferencing (Huth et al., 2017; Schiliro & Choo, 2017). The scope of ICT is much broader to Information Technology (IT) which is often considered to be a subset of ICT as it envelopes both internet and mobile-enabled wireless networks comprising of obsolete ones such as landlines, radio and telecommunication transmissions which in today's world are used along with modern bits of ICT such as robotics and AI. Some of the recent entrants of ICT include digital TVs, smartphones and robots and older ones as computers and telephones continue to advance and used for decades now. There are certain components (Europeyou Association, 2020) that makeup ICT as shown in Figure 7.1. The role and the importance of each component are given as under.

7.2.1 HARDWARE

It refers to the tangible physical components that make up an IT system that may include an electronic or computer system and everything that has been assembled to make it such as CPU, memory, hard disk, monitor, CD drive, etc. It goes hand-in-hand

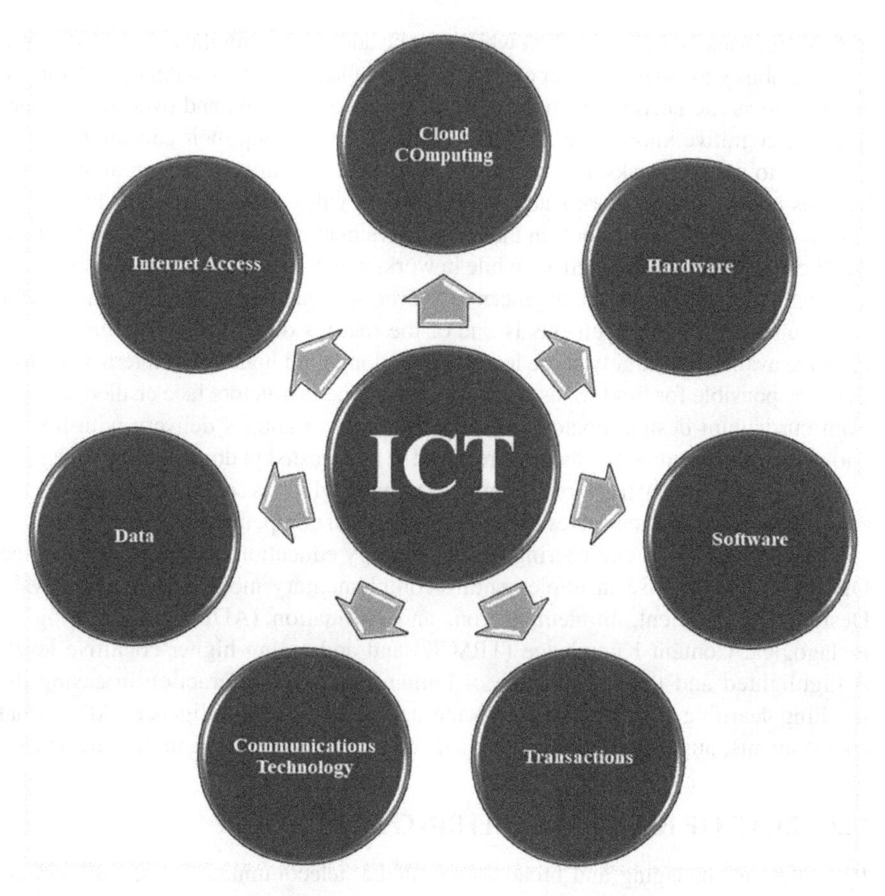

FIGURE 7.1 Components of ICT.

with other necessary components such as firmware and software for functioning of the IT system.

7.2.2 Software

It is a set of instructions, data or programs employed to function computer systems and perform specific tasks. This refers to scripts, programs and applications that function on a device in the form of a scripted program. It is variable in nature and is often split into application or program furnishing a need or want, downloaded by users and a system which comprises of operating systems and programs supporting applications.

7.2.3 Transactions

Transactions (digital) are automated/online paper-free communications that happen between organizations and people. They are important as they result in saving time and money along with providing enhanced customer satisfaction by reducing the need of physical presence of face-to-face interaction and hassles involved in physically performing the activity.

7.2.4 Communication Technology

It refers to the consortium of program and equipment that is employed to handle, manage, process and transfer information. It involves the development, installation and troubleshooting of the hardware and software systems. The new day trends involve the use of microelectronics in communication technology for diminution, customization, automation and globalization for faster, cheaper and reliable communication to a wider audience.

7.2.5 Data

It signifies information (digital) using machine language systems that can be construed using various technologies involved in the same. The binary system is capable to store video, text or audio information in bits, i.e. 0's and 1's of <<on>> or <<off>> values. A four-digit pin number or debit card transactions record can be termed as data.

7.2.6 Internet Access

It involves users/enterprises getting connected to the internet via devices such as PCs, laptops or cell phones. There can be varying data speeds on offer based on the region-wise data subscription packs, service provider and type of connection for availing web-based services.

7.2.7 Cloud Computing

In general, cloud computing is a term employed to label data centers that are available to multiple users across the internet via the use of central servers. These may be of an enterprise (limited to one organization/enterprise), public (available to many organizations) or hybrid nature (a combination of both). If the user's connection is close, it may be labeled as an edge server.

With the virtue of penetrating all spheres of human life, ICT has seen tremendous growth across the world. The ICT industry is rapidly growing with respect to annual rate of growth in market share, production, investment export and offshore outsourcing (Jain & Agrawal, 2007). The recent literature suggests the use and consequent growth of ICT in education (Mlambo et al., 2020), medical sector (Stadin et al., 2021), energy sector (Lange et al., 2020), industry sustenance (Abramova & Grishchenko, 2020), banking sector (Del Gaudio et al., 2020) and employment (Asongu & Odhiambo, 2020) to name a few.

ICT has become an important tool for simplification of human work. The field of education is one of the areas which have fetched successful results with the use of ICT for some decades now. The incorporation of ICT in the teaching–learning process has allowed the teaching community to shuffle from the traditional classroom methodology and reshape concepts and strategies for teaching–learning to supplement classroom activities, rearranging course teaching structures, thereby offering more autonomy to learners so as to make learning happen from their perspective. It has enabled teachers who were already on the lookout for modern technological innovations for uplifting the quality of their teaching.

Table 7.1 gives an account of changes brought about by the use of ICT in traditional engineering teaching learning methodology.

It is fair to say that technology is advancing at a pace and it is essential for the teaching–learning community to keep abreast with the same. ICT enables easy access and updating of learning materials along with a library of additional teaching materials like research journals, articles and online books on offer. The learners choice of pace for learning along with the time, and setting for study, provision of instant system feedback and stimulation/motivation due to gamification feature that can be enabled in the activities/tasks makes it student-centric in terms of learning. This modern way of teaching can occur with teacher not being physically/virtually present during the learning activity. Also, more discussion, teamwork, dynamism in quick and real-time assessment through assignment and its evaluation and ease in distance learning combine, to sum up, the advantages ICT offers to engineering education as a whole (Bhattacharjee & Deb, 2016). However, there are some courses such as language, the learning of which might be better with personal contact in a conventional teaching setup. Also, the use of technology poses problems in terms of shortage for competent teachers with respect to use of technology as they might have spent all

TABLE 7.1
Traditional teaching methodology in comparison with modern teaching approach with use of ICT in engineering education

Serial No	Traditional approach of teaching for engineering courses	Modern approach with ICT in teaching for engineering courses
1	Learning is teacher centered	Learning is student centered
2	Mass instructional behavior of teaching where pace is same for all	Mass customization fitting instructional learning needs of individual with flexible pace
3	Learning limited to classroom during school hours	Learning is possible outside class room and school hours
4	Largely limited to learning of facts and recitation	Approach higher forms of learning such as critical thinking and problem-solving in real world context
5	Student performs individually	Collaboration and discussion among student/s and/or teachers
6	Textbooks, the major source of knowledge, can't be updated with ease	Information resources (online) are up to date and can be easily updated to meet students demands
7	Progress discussion meetings scheduled once/twice per semester	Communication regarding student progress is possible on a real-time basis

their life teaching with the traditional classroom methodology which majorly emphasizes content delivery and ICT might be time-consuming with content creation and its preparatory requirements. There are engineering establishments in the modern world which are deprived of basic ICT facilities due to insufficient investments. Furthermore, the same is applicable even to learners who might have limitations in resources such as smart devices, access to technology/internet, etc. At times, the feedback might also be inconclusive due to the limited use of ICT tool by a learner and most importantly emotional learning is absent which is an integral part of human nature (Klimov, 2012). The intermittent change of curriculum, focus on mass education rather than the quality, lack of association between academics and industries, lack of resourcefulness from the corporate sector, generous assessment for quality accreditation and dependency nature in an institutional atmosphere are some of the other reasons for ICT not being used up to its potential (Jain & Agrawal, 2007).

7.3 OUTCOME-BASED EDUCATION

"Does teacher-centered education really support innovation?" This question can be considered as a triggering point for the need of initiating reforms in the context of engineering education system and the allied areas in educational terrain. The traditional teacher-centered education process has sustained over a long period of time as there were various socio-cultural, modernization and industrialization aspects responsible for defining traditional education in between the 19th and 21st century. The motive of traditional education was to transmit the facts, skills, along with the set of moral and social conduct that earlier generations considered to be necessary for the next generation's sustainable and holistic growth. The traditional approach had its own features like the classroom being orderly and students being passive while the teacher retained full control sufficing the greed of transferring the content at a mass scale. John Dewey (Dewey et al., 1904) described the next generation in this traditional setup as "imposed from above and from outside". The next generation was expected to be docile and obedient in receiving and believing the fixed answers. The teachers were considered as instruments through which the knowledge was communicated and the standards of behavior were enforced. The traditional pedagogy was inadequate for the acquisition of facts and figures about what and how much knowledge was gained by the students, and whether or not a student knows and can justify the credentials earned. If the process of teaching and learning is teacher dominated, instruction often becomes boring for students resulting in missing the essence of learning. Also, the students were not allowed to express themselves and direct their own learning (Otukile-Mongwaketse, 2018; TABULAWA, 2003). The traditional teacher-centered pedagogy is associated with top-down, hierarchical pedagogy, and reinforce passive learning, rote memorization with little to no effort at understanding the meaning of content and hindered the development of higher-level cognitive skills (Cristillo et al., 2016; Daher, 2012). These pedagogies are also associated with authoritarian, anti-democratic regimes that exert centralized control over schooling to produce an obedient passive citizenry. The graduates lack the required level of awareness about technical skills and

soft skills for the contemporary job requirements and are not adequately prepared to become workforce for the industry.

There was a need to overcome the inertia that these educational institutes had, as a result of which they followed older conventional practices and structures. Even though these practices were unproductive, they were adopted due to their familiarity and comfortable nature (Spady, 1997). With this scenario, the education reform movement adopted progressive education techniques as a new educational theory, i.e. OBE, which focuses on skills and results achieved by the student as the most important aspect of education. This education reform was formally framed and accepted as the Washington Accord. The Washington accord is a recognition of the substantial equivalency of accreditation systems of member states and permits the graduates from member states to work in other member states. OBE finds the mark/grade-based educational system irrelevant and does not rely on conventional teaching methods. It believes in employing assessments, opportunities, classroom experiences and provision of all the necessary support to the students for achieving their goals instead of targeting rankings through examinations. The students perform better due to the indirect instruction methods wherein the central focus is to enhance student skills (Kratochwill et al., 2000). W.G. Spady compiled all the aspects of OBE and proposed that the main focus of OBE in an educational system is on the student and the abilities they acquire at the end of a given learning experience. To ensure the results, it is the responsibility of the institutes to start with a clear vision of what students have to achieve and subsequent arrangement of instruction, curriculum and most important assessment to enable the ultimate goal of learning (Spady, 1994). The main difference between the philosophy of OBE and traditional education is that it primarily focuses on outputs or outcomes as products that can be measured and assessed rather than on the resources that are available to the student, which are called inputs.

OBE is primarily a performance-based approach with the student being at the center of the teaching–learning process. OBE in itself does not specify or require any particular style of teaching or learning. On the contrary, it requires the students to demonstrate the skills and content learned. It is believed that all students can learn, regardless of their ability, race, ethnicity, socio-economic status and gender. Furthermore, a complex organization, if provided with the framework, is capable of producing what it measures and to downplay anything it considers unimportant. The adoption of measurable standards is seen as a means of ensuring that the content and skills covered by the standards will have a high priority in student education. The role of teacher, graduate attributes and, most importantly, student participation toward the development of a good engineer is depicted in Figure 7.2.

In OBE, the desired outcome is selected first and the curriculum, instructional materials and assessments are created to achieve the intended outcomes. All educational decisions are to be made so as to facilitate the desired outcomes. In this evolved approach to planning, delivering and evaluating instruction, the administrators, teachers and students are required to focus their attention and efforts on the desired results of education that are expressed in terms of individual student learning. It can be concluded reasonably (Killen, 2000; Spady, 1994) that the philosophy of OBE stands on the following considerations:

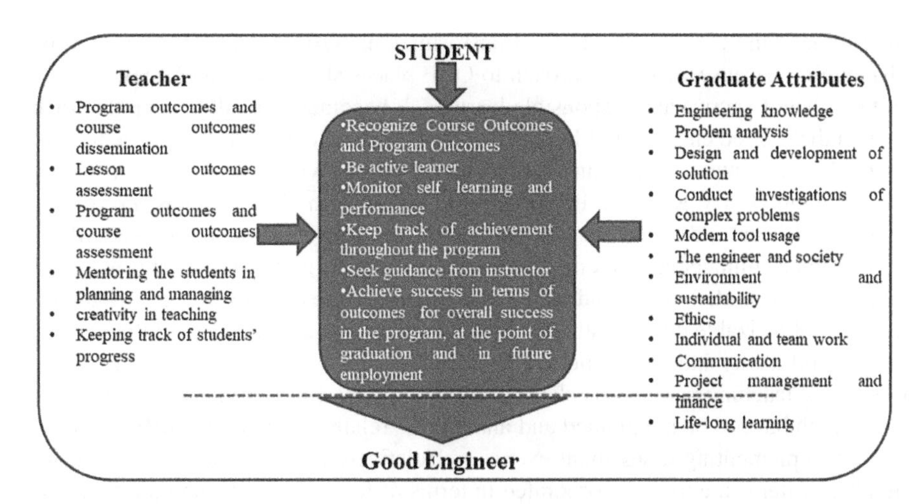

FIGURE 7.2 The guideline for the framework of individual roles of teacher, student and graduate attributes to produce a good engineer.

a) All learners have the competencies to learn and succeed, but not all at the same time or in the same way.
b) Successful learning promotes even more successful learning.
c) The institutes and instructors control the conditions that determine whether the students are successful or not.
d) The role of instruction is to find ways for students to succeed, rather than fail.
e) All institutes which follow OBE are driven by mutual trust.
f) Excellence is for every student and not just a few of them.
g) If the students are prepared every day for success, it reduces the further need for corrective measures.
h) Learning is motivated by collaboration and not competition.
i) As far as possible, no learner should be excluded from any activity in an instructional environment. If provided with opportunities, every student will participate in the learning process.

The philosophy of OBE is designed to improve student learning and have various advantages over the traditional education system as discussed above. However, there are concerns about the grassroot level implementation of OBE defining the doctrine and technical support required. The main concern of the traditional learning model is that the assessment of learning becomes the last step of the learning process and the immediate course objectives, competencies expected at course level are significantly negligible, making it difficult to converge with a broader program or institutional objectives (Andrich, 2002; Tuckman, 1988). The inherent time defined structuring of OBE demands for real-time response and continuous active participation at both ends of the teaching–learning spectrum. On the other hand, adjusting OBE methods within the time-defined structure of traditional curriculums and delivery mechanisms may also limit the effectiveness of the method (Spady & Marshall, 1991). In contrast to the traditional model, the implementation of OBE demands very high student involvement and participation in the transformational OBE framework. This can be challenging because such an approach to OBE places students in the center, thereby turning them to active and responsible learners. It becomes difficult to keep the track toward the desired end-point of learning as the course outcomes are linked to the program objectives and in turn connected to objectives across other disciplines providing a continuum (Barry Kissane, 1995). A lot of features in OBE systems, upfront or behind the curtain, demand the content transfer to be done outside the class which may become cumbersome as it modifies the role of a teacher from content provider to knowledge presenter, mentor, advisor and facilitator of the learning process. The management of a sizable class of students with involvement of so many activities, maintaining uniformity and reach among all students, to go along with ensuring the quality of creative teaching practices is also a point of concern (Kerr, 2011). The data generated which have to be maintained and analyzed is relatively higher in OBE as setting up and implementing assessment events are relatively more frequent and a greater number of items are to be incorporated in terms of formative and summative assessment tools. There are studies available reporting traditional forms of assessment along with standard problem-solving etc. being simple and easy in terms of examinations whereas newer forms of assessment that offer an authentic representation of practice could be more complex and expensive (Palmer, 2007). It was further reported that the scope of assessment goes beyond understanding it in the form of communication, primarily between student and teacher and in a wider perspective being a communication link between the education system and wider society encompassing various stakeholders such as employers, curriculum designers and policymakers (McAlpine et al., 2002). The assessments varied in new ways such as time, context, test design, performance standard, linking of assessment items, etc. are the features beneficial in building exhaustive mechanisms for testing the skills of students. This, in turn, contributes toward OBE which is not easily possible to implement without the use of ICT tools. However, there is a scope to surpass the limitations of traditional assessment systems and approaches with the incorporation of ICT-enriched assessment methods as alternative assessment approaches and instruments for measuring the complex, higher-order outcomes. These assessments designed with ICT bring new understandings that have the potential to impact learning, assessment and overall society at large. ICT possesses the potential to play a major role in transforming assessment practices to support the needs of learners as well as the needs of educational systems.

7.4 ICT TOOLS FOR AUGMENTING OBE

In today's globalized world, there is a constant need for adapting to dynamic changes happening around us. Apart from the knowledge of core engineering subjects, future engineering graduates also need to acquire a new set of skills and competencies for applying the knowledge in order to bring about the change. The implementation of the OBE framework for engineering education can ensure the preparation of students for which three crucial processes regarding teaching practices need to be fundamentally changed. These processes include curriculum design, i.e. what to teach; education delivery, i.e. how to teach; and student assessment/evaluation, i.e. how to assess student learning.

The traditional student assessment and evaluation model is largely focused on examinations which mainly involve the summative assessment. The quality of examinations in engineering education is often criticized of mainly being based upon the memorization of concepts rather than evoking a higher level of thinking. The assessment process based on rote learning doesn't leave any scope for assessment of higher-order cognitive, psychomotor and affective abilities. A reform in the examination system is essential as the traditional education system is focused on the measurement of "how well students perform in the examination" rather than "whether the desired learning outcomes have been achieved or not" (Mecwan et al., 2015). This form of examination even influences the teaching methodology of teacher, i.e. content delivery and the student's attitude toward learning (Preston et al., 2020). The students are mainly focused on earning the grades rather than acquiring and applying the knowledge and skills. It is reported that "assessment drives learning" (Cox et al., 2014) and hence a change in the assessment system can be a catalyst for bringing changes to the curriculum design and delivery aspect of teaching.

The graduate attributes, often referred to as Program Outcomes (POs), are the skills, abilities and attributes engineering students should develop toward the end of the program. POs are the common attributes to which all the stakeholders, i.e. student, teacher, industry, parents and alumni, agree. The framework of Bloom's taxonomy is often prescribed to assess these outcomes. It can be achieved by aligning the assessment items with the desired student learning outcomes (Kastberg, 2003). Although POs represent the bigger picture, describe broad aspects of knowledge, skill and behavior, they are not specific enough to be measured directly. Each PO is further divided into a set of competencies that are more specific and tangible as compared to the POs. The performance indicators (PIs) are constructed as specific and measurable statements of student's expectations for each of these identified competencies. These are the "leading indicators" of appropriate achievement level or competency defining the acceptable level of proficiency (McCahan & Romkey, 2012). To measure the specific learning outcome, flexibility is provided to mold the PI as per the course outcomes. In a nutshell, they act as measuring tools in the assessment of competencies and, in turn, competencies will act as a measuring tool for the assessment of POs. While designing the competencies and PIs, the "design down" approach is selected whereas, for assessment, the reverse approach is adopted as shown in Figure 7.3. With this approach, it is possible to link each assessment item and opportunity with PI.

In the present examination system, assessment items and opportunities are mostly limited to evaluation of recall/understand level of factual knowledge, which is only a little part of several abilities that are expected to be demonstrated by the engineering graduates. The assessment process must also be capable to test higher level skills, like ability to apply knowledge, solve complex engineering problems, analyze, synthesize and design. It is essential to attain professional skills like the ability to communicate, work in teams and ability to inculcate lifelong learning as they have become important elements for the employability of the graduates (Shuman et al., 2005). The use of the summative assessment approach is employed conveniently for the assessment in engineering education. When it comes to the assessment of higher level and professional skills, it becomes difficult to use direct measurement tools and they are usually evaluated on the basis of judgment and experience. Some of the

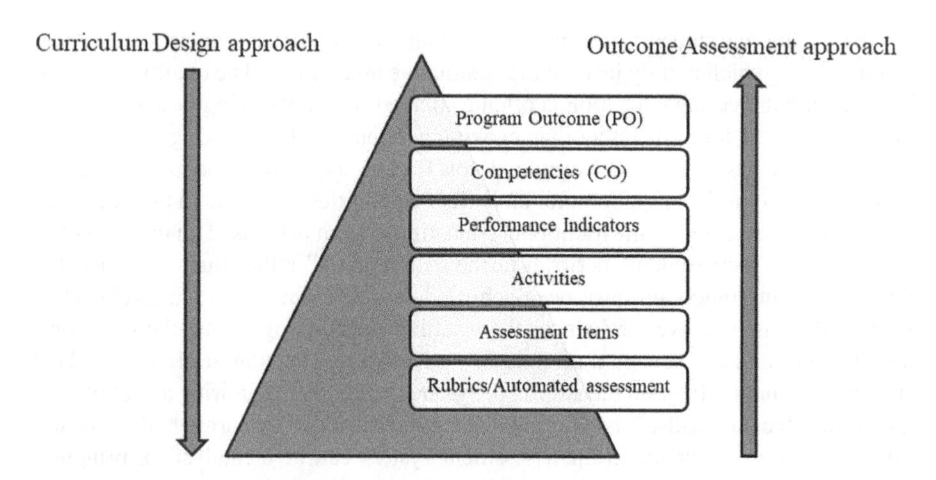

FIGURE 7.3 Outcome assessment approach for OBE using ICT.

reliable assessment tools/methods such as rubrics ensure accuracy and uniformity. These are largely employed for assessment and grading as a scoring framework for instructors which will help to evaluate the performance of a student for specific activity based on several performance criteria, for which rating is done on a PI-based scale.

However, designing such a complex evaluation system and executing it needs a lot of skills and efforts along with an enormous amount of work. Figure 7.4 shows a tree diagram representing the approach that has to be taken from POs to assessment items. Let us consider an undergraduate engineering course/program to understand the magnitude of work required to be done. An undergraduate course can be mapped to six POs with each subdivided into four competencies. These competencies are further divided into four PIs each and two assessment items are used for each PI. This minimalistic approach would still result in the design of 192 assessment items. Even if designing could be considered as a one-time activity, assessment and evaluation of

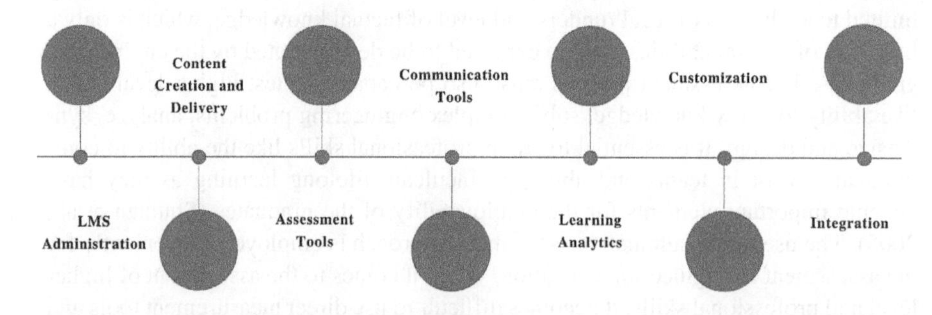

FIGURE 7.4 Assessment of learning outcomes in OBE.

each student for a number of courses will have to be done. These tasks in addition to the evaluation of outcomes at the course and program level will result in a huge amount of work. The consideration of design as a onetime activity is keeping in mind that the curriculum remains the same. However, there is a frequent change in the engineering curriculum these days due to the need of adapting to the ever-changing technology. ICT can play a major role in the design and execution of evaluation systems for the implementation of OBE in engineering education. Also, LMS can be of great help in designing and executing such a complex process (Crespo, Najjar, Derntl, Leony, Neumann, Oberhuemer, Totschnig, Simon, Gutiérrez, & Kloos, 2010). It will largely benefit to the teaching community as their focus can remain on the teaching–learning aspects rather than assessment and evaluation of outcomes which otherwise involves a lot of manual efforts.

7.4.1 Learning Management System

Virtual Learning Environments (VLE) is a web-based system for the delivery of learning content to the user. It is becoming an important tool for the education in engineering and technology domain. An LMS is a software that is designed specifically to create, distribute and manage educational content along with assessment and evaluation of student learning. LMS generally provides a web-based interface that supports a wide range of activities which includes but is not limited to media resources, digital content, forums, questionnaires, chats, assignment, file sharing, etc. There are several open-source LMSs available like Moodle, Blackboard, ATutor, Open edX, Chamilo, Canvas, Forma, Dokeos, Ilias, etc. and commercial LMS options like Litmos, LearnUpon, TalentLMS, Google Classroom, etc. (Cavus & Zabadi, 2014). Some of the commercial LMSs like Google Classroom, offer free services to some extent and are becoming popular among teachers.

With the presence of several LMS options in the market, choosing a right LMS which caters to the teaching–learning needs of the stakeholders involved in the exchange is very important. The various features of LMS are illustrated in Figure 7.5, which can be looked upon as a part of the process to decide the best suitable option.

7.4.1.1 LMS Administration

LMS administrative tools decide its overall operational performance. The administrators have to perform various tasks like the setting of the learning path for users,

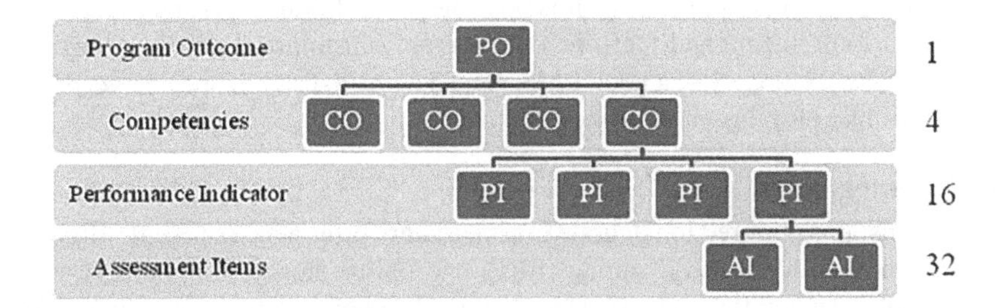

FIGURE 7.5 Features of a Learning Management System (LMS).

creating and managing courses, creation of users, roles and authentications, uploading content and many more.

7.4.1.2 Content Creation and Delivery

Content Management System (CMS) is an integral part of any LMS as it is responsible for the creation, storage and delivery of digital learning content. A good LMS supports all possible type of contents essential for e-learning (pages, text, links, images, videos, etc.) and provides scope for adding a new type of content. The success of LMS from a teaching perspective is largely defined by ease and flexibility in content creation and delivery.

7.4.1.3 Assessment Tools

Apart from typical assessment tools like tests, quiz and assignments, LMS offers a large number of assessment features such as automated testing and evaluation. The ease and flexibility in the creation of assessment items and reliable execution are the key parameters of an LMS. To measure the learning outcomes, it is essential to create various assessment opportunities within an LMS that otherwise is not easily possible without its use.

7.4.1.4 Communication Tools

LMS offers various communication channels for instructors and learners like e-mail, message, discussion forum, live chat, video conferencing and whiteboard. These tools are useful for formal and informal communication within the learners along with instructors–learners setup. It is an essential quality of good LMS that it should provide and promote these communication channels which are critical for the success of a learning program.

7.4.1.5 Learning Analytics

Learning analytics (LA) is a methodology of collection, measurement and analysis of learners and learning data, which can be used for understanding and optimizing learning experiences and its impact at the institute or university level. LA is helpful for tracking the learner's progress throughout the journey of learning path, with a provision of personalized feedback to the learners.

7.4.1.6 Customization

In general, LMS offers the flexibility to modify its features and functionalities as per the teaching–learning requirements. It is possible to modify the user interface layout (customized look, theme and feel), report structure (customized reports), language and many more features with a good LMS. It also helps to cater to specific needs and add features like plug-ins or add-ons.

7.4.1.7 Integration

On occasions, it is required to integrate other software programs like Single-Sign-On (SSO), Enterprise Resource Planning (ERP) and online storage programs (Google Drive, Dropbox, iCloud) with LMS. However, the level of such integration with LMS has to be verified on case by case basis for enabling this feature.

TABLE 7.2
Comparison between open-source and commercial Learning Management Systems

Open Source LMSs	Commercial LMSs
Do not charge license fees	License has to be purchased
The requirements in terms of time and efforts for implementation of such systems are high	Ease in relation to time and efforts for implementation of such systems
Hosting requires in-house expertise, hardware and training facilities are required	Minimal nodal requirement as it comes embedded as a package
Customization for individual/institutional teaching–learning requires expertise	Customization can be discussed before purchase and embedded as a package for individual/institutional
Internal support/maintenance system required	Comes as a part of the package as per the licensing agreement

There are certain advantages of using open source LMS as they are free, scalable, with the possibility of customization through add-ons/plug-ins along with the presence of an active online user community. However, there are limitations in terms of not being user friendly and unavailability of technical support. The criterion that may be used as a part of the selection process of an LMS on the basis of the licensing aspect (free/commercial) is given in Table 7.2.

Blackboard, a LMS developed by Blackboard Inc., is web-based software which offers features like virtual classrooms, course content management, communication tools, collaboration platform, customizable open architecture, scalable design and assessments/evaluations system. Blackboard Analytics is a LA tool that allows the monitoring of the student's progress and their levels of online course engagement over time. It provides detailed analytics like time spent within the Blackboard site, the total number of clicks on content items and total time taken to submit work by the student which is both interesting and useful. Google Classroom is a LMS that is a part of Google's *"G Suite"* for education. It provides free services for schools, non-profit organizations and personal use. The main feature of Google Classroom is its integration with existing Google products such as Gmail, Google Drive, Google Calendar, Google Forms and other popular applications, which makes it very easy and convenient to use. Moodle is the most widely used open-source LMS which can be effectively used for content delivery, design, execution of assessment and evaluation system in OBE (Barge & Londhe, 2014). Moodle is an internationally accepted LMS which is has a strong community base. It provides ease in implementation and supported by the best architecture with inherent customization and integration options (Al-Ajlan, 2012). The use of Moodle in engineering curriculum for content delivery and assessment is well studied in literature and significant impact on various aspects of the teaching/learning process and student performance is reported (Kulkarni et al., 2019). Furthermore, in recent years, Massive Open Online Courses (MOOCs) offered by edX have brought transformation in online learning. edX is a MOOC platform, created by Harvard University and Massachusetts Institute of Technology, which is supported by more than 150 educational institutes in terms of content creation, with more than 3000 online courses offered to accommodate 33 million learners across the globe (Shah, 2018). The source code of edX-based LMS is available on GitHub as it is an open-software tool. One of the important reasons for wide acceptance of all these LMSs among learners is the provision of mobile applications for each of Blackboard, Google Classroom and Moodle, with which learning content can be made accessible through mobile application on Android and iOS mobile devices.

7.4.2 New Developments in LMSs Inclined Toward Augmentation of OBE

LA tools of LMS hold the potential to access the learning outcomes of the students. LA provides qualitative and quantitative data about the student's progress over the learning path. This data can be further used for the assessment of learning outcomes and valuable feedback can be provided to the students for improvement. With the advancements in LA, it is possible to provide personalized learning modules, wherein course content, course delivery and assessment can be optimized as per the need of an individual student. Moodle offers many useful analytical tools that can be adopted, integrated and customized to develop a system for assessment of the students learning outcomes (Yassine et al., 2016). It is possible with Moodle learning analytics to provide personalized learning support via the use of activity-based smart indicators for tracking students' progress and provide them with recommendations for improvements (Florian et al., 2011).

Hu et al. implemented LA dashboard using Moodle and Open edX for monitoring the student progress of outcome-based learning at a university in Hong Kong (Hu et al. 2017). The course contents and activities (page, text, video, quiz, etc.) were linked to course outcomes using HTML block page in Moodle and XBlock in Open edX. As the student progresses through various learning contents and activities in Moodle and Open edX, the performance of each student activities was measured in terms of course outcomes in real time. The learning outcomes were evaluated at the end by aggregating the scores of all the components linked to a specific learning outcome which made the process effective and assessment was done in real time.

Moodle 3.1 version and above provides the features to create a competency framework for evaluation of student performance. It is further possible to breakdown competencies into PIs which can be mapped to various activities in Moodle at the level of assessment items. While designing a course, one can make sure that there are activities or resources for various competencies and each of these can be linked to various activities. Rezugi et al. proposed an ontology-based competency management application as a Moodle extension (Rezgui et al., 2014). The plug-ins or extensions work as customization tools that enable you to add additional features and functionality to Moodle as per your requirements.

Rubrics is a powerful tool for assessment and grading of student work and the use of data mining techniques helps to trace user interaction with learning resources, platforms and student–teacher interaction. Rayon et al. proposed Scalable Competence

Assessment through LA (SCALA), an approach for the improvement of competency-based assessment along with the use of enriched rubrics and data mining techniques (Rayón et al., 2014). This system provides an easy-to-use dashboard with rubrics and learning metrics wherein teachers can monitor student's progress throughout the learning path.

Moodle can be integrated with other useful tools while teaching an engineering course. The development of a Moodle module called "CTPracticals" was done for performing the automatic assessment of laboratory work entitled, "The design of a basic CPU" in the course of Computer Technology (Gutiérrez et al., 2010). In this module, the students work collaboratively on a design assignment using an online platform. The students could ascertain the state of their work and after the submission can access the result of automatic assessment along with feedback from their teachers. Furthermore, teachers get the statistics of overall class status and can provide feedback on an individual basis. This partially relives teachers from the cumbersome activity of submission assessment, and they can focus more on the teaching–learning process.

The professional skills and competencies like "Commitment" and "Teamwork" is very intangible in nature. In general, the teacher accesses these skills on the basis of personal judgments and biases. There have been attempts to verify whether LA approach in Moodle could be suitable to find relevant predictors for such competencies among the students. However, such efforts are still at an infant stage and one can expect the use of Moodle analytics for such assessment (Iglesias-Pradas et al., 2015). There are also attempts made toward the development of Moodle plug-in which will evaluate student skills in thinking and innovation (Chootongchai & Songkram, 2018).

An intelligent tutoring system (ITS) is a computer-based system that provides one-on-one immediate and customized instruction with identification of learner's knowledge gap with a provision of giving feedback to learners in absence of human interaction. It is possible to develop such a system using Moodle (Ramesh et al., 2015). ITS will be discussed in detail in further sections.

MOOCs are online web-based courses offered by various institutes and universities across the world. They have received a massive response in the last few years as most of them are free for enrolment and enable learning without the limitations of time and location with edX, Coursera, Udacity, Udemy as some of the key players offering them. Initially, most of the MOOCs were offered as a separate course and were not part of any formal engineering program. However, now they are becoming more structured and are offered as a "mini" and "micro" degree program and are becoming a part of higher education (Sandeen, 2013). Some colleges and universities are integrating MOOCs as a part of their credit system (Cui & Wang, 2017). Yanshan University, China, has presented an interesting study where MOOC was developed based on the OBE model for "Fundamentals of engineering drawing" with "XuetangX" an online free and open MOOC platform and offered the course to more than 1200 students (Jian-Feng et al., 2017). The course reported herein has been offered on four occasions since March 2018 successfully as reported on the XuetangX Platform (Song et al., 2020).

7.5 ICT'S USE IN ENGINEERING EDUCATION

In the context of engineering education, the framework provided by OBE demands teaching–learning activities like content creation, management and organization of knowledge, assessments, setting up various training activities, student performance tracking and skill set training to be effective. This has to be done to ensure that the process is more rigorous, effectively implemented and efficient in terms of record keeping. As discussed in the earlier section, one may face many challenges in the process to achieve this goal. The role of the teaching community to impart OBE in a true sense goes beyond teaching–learning activities and its analytics. Furthermore, it pushes for a holistic approach of instructional design methods along with the requirement for the categorization and organization of the tutelage in all three contexts: design, delivery and assessment.

The decision-making in the selection of instructional methods is becoming more complex with the increase at both the fronts of demands and options available. Instructional design is the designing of learning events based on instructional development models. It is the teaching instrument that makes instruction as well as instructional material more efficient, effective and engaging. The lack of training in instructional design may lead to such important decisions being taken on the basis of convenience, comfort and/or trends. There are several such instructional design approaches to explain the design and development process for any learning situation and content such as First principles of instruction by M David Merrill (Merrill, 2002), Bloom's Taxonomy (David & Krathwohl, 2001) and Gagné's Nine Events of Instruction (Gagné et al., 1992), etc. Such models help educators to manage learning units for the most effective learning environments. The ADDIE instructional design model is reportedly useful, dynamic and flexible for creating and managing learning objects such as workflow designing, implementation and instructional designer's primary considerations (Melanie McGurr, 2008; Ozdilek & Robeck, 2009). ADDIE model comprises of five clearly defined phases namely Analysis, Design, Development, Implementation, and Evaluation (Seels & Richey, 1994). The sequence in phases does not impose a strict linear progression through the process. The positive impact on LO with the implementation of ADDIE model in the instructional design of "Structural Mechanics" course as the core curriculum of civil and hydraulic engineering was reported to have higher achievements of learning outcomes in the designed course (Xing, 2018). The use of ICT in developing online project-based collaborative learning during the prototype phase using ADDIE for soft skills and competency testing to cope with employer's requirements is reported (Nadiyah & Faaizah, 2015). A study of project-based software engineering course besides having well-founded pedagogical design and implementation, with an extension of capstone projects, is also carried out (Lewis & Smith, 1993). The role of ADDIE in the development of an open online course for the professional development of teachers with consideration to the diverse needs of the participants is reportedly discussed (Trust & Pektas, 2018). It is quite evident from the scholarly works cited here that as an instructional design model, ADDIE, which facilitates implementation of effective training tools, has found widespread acceptance and usability across various engineering streams among teachers, researchers, instructional designers and training developers.

TPACK is a set of knowledge encompassing the content that needs to be taught to the students to enable effective teaching using technology. It is an integration framework that identifies technological, pedagogical and content knowledge as three sections that need to be combined by the teacher. TPACK in its abbreviated form stands for three components:

1) Technological Knowledge, which is about the use and integration of technology in teaching. Technological knowledge is the know-how, understanding and use of tools that need to be incorporated in teaching.
2) Pedagogical Knowledge, which refers to the art and practice of teaching with effectiveness. It is the knowledge of "how to teach" in order to make the students learn in the best possible way and putting it into the practice of the teaching process for knowledge transmission.
3) Content Knowledge, which is the teacher's knowledge about the course content that would include knowledge of concepts, ideas, theories, experiments and frameworks.

TPACK is an important dimension to explore for the topic under consideration as it incorporates the relevant use of technology in the classroom as well as continuing the focus on the content and the way to teach. It is important for the teacher to be completely up to date and knowledgeable with the curriculum and the components of TPACK to effectively incorporate it into their teaching. The present-time engineering students are competent to learn better with the incorporation of technology and often the direct teaching methods involving conventional content transfer do not appear to be appealing. The modification of introducing the technology component into the conventional pedagogical knowledge content model makes the students become more engaged in their learning.

The "technological components" in TPACK model comprise of the technologies available and training required to use the technological tools along with the associated resources. It includes LMS, modern gadgets and interfaces, cloud computing, multimedia technologies, 3D printing, high-end virtual technologies, augmented technologies and LA technologies. These technological developments have made it possible to address individual needs thereby easing and driving the decision-making process. It enables the students to learn with more tangible and physical experience and allows access to information stored in remote cloud servers for accessing the coursework and educational materials from any device with anywhere and anytime approach. ICT tools such as computers, LCD projectors, overhead projectors, television, mobile phone, interactive whiteboard and application software contribute to the "pedagogical section" of TPACK. The spread of ICT also contributes to "content knowledge" and brings in an interesting dimension for teaching communities about overall awareness of recent developments in the topic they are about to teach. The various ways in which the content could be presented helps to gain an insight in terms of understanding of the content. The availability of content and expressing the content in various possible ways among the teaching community has been made possible with the use of ICT.

The understanding of the TPACK model and its role in imparting technical and engineering education from the teacher's perspective is explored in the literature (Mutanga et al., 2018). It was reported that most of the teachers participating in this study were using ICT in their teaching–learning methodology and opined that the implementation of TPACK had a positive impact on the quality of technical and engineering education. The role and importance of design thinking, validation and subset of technology and pedagogy in terms of knowledge in engineering and allied streams was exhaustively reviewed and emphasized (Chai et al., 2013). A study was carried out wherein it was accentuated that the consideration of TPACK by the teaching community is much needed to develop student competencies and for developing ICT-integrated lessons (Koh et al., 2015).

The definition of higher-order thinking was attempted and the confusion with allied terms is removed by setting up a larger platform for discussion. It was commented that a teacher needs capabilities to assess student's higher order thinking skills (Lewis & Smith, 1993). It was further reported that ICT-enabled classroom environments had a positive effect on the acquisition of student's higher-order thinking skills. Several implications related to classroom design to enhance the development of higher-order thinking skills were studied and it was opined that ICT-enabled classroom environment was significantly different from the traditional classroom (Hopson et al., 2001). It is quite evident that ICT has a significant role in achieving higher-order thinking.

ICT offers features like ease of accessibility, usage and availability along with online–offline availability of resources. Furthermore, it ensures the replication of the current setup with providing a faster deployment rate to play a vital role in the use of technology for improving the teaching–learning process. In addition, the technical features allow customization to meet various demands of an institution. The use of ICT for the effective teaching–learning process is reviewed by many researchers for various considerations like enhancement of teaching–learning (Majumdar, 2006), increased individual capabilities and personal experiences of students (Shahmir et al., 2011), various issues surfacing with the implementation of ICT (John & Sutherland, 2004), a general model for integration of ICT into teaching and learning (Wang, 2008) and overall TPACK-ICT integration (Malik et al., 2019).

The three independent elements, i.e. OBE, Instructional design and TPACK integration model converge, and it may be hypothesized that the aspects of ADDIE and TPACK in terms of engineering education by the community and its implementation with ICT's aid would greatly help the teaching community in achieving the targets set by OBE. This complementary behavioral relationship between ADDIE and TPACK because of their own nature and their contribution toward articulating OBE with the help of ICT is shown in Figure 7.6.

7.5.1 Role of Community Building and Student Participation

The advances in ICT have given us an opportunity to communicate and collaborate with literally anyone in the world which was earlier not possible. The online communities are taking over as prominent platforms for the exchange of ideas and resources by forming a "Virtual Platform." These platforms represent an informal

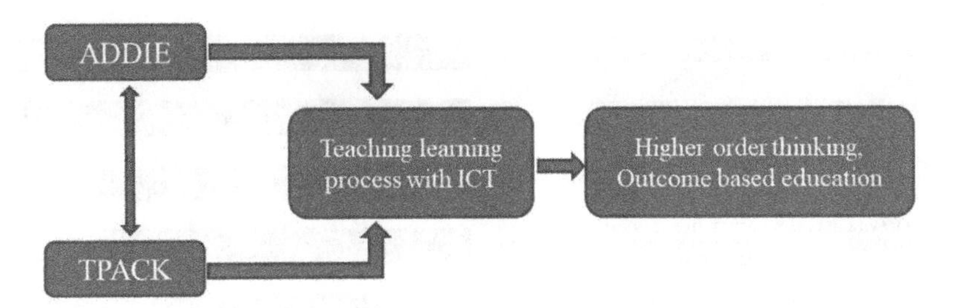

FIGURE 7.6 Role and interplay of ADDIE and TPACK in the teaching–learning process with incorporation of ICT tools for effective implementation of OBE.

communication channel wherein people across the globe get connected for a specific purpose with each other via social networking sites, emails, forums, chats or blogs. There are many communities such as open-source software communities that go beyond just sharing by collaborating virtually for the creation of recourses. The successful implementation of OBE requires the contribution of various stakeholders like educational institutes, students, parents, industry personnel and the community at large. All these stakeholders must come together to share the ideas, resources and collaboration for implementing OBE.

To make the most of the features of an ICT system, active participation from both the student and teacher side is essential. The student is considered to be at the center of OBE as its focus is on analyzing the conceptual understanding of the student and customizing the methodology of imparting knowledge to them as per their individual requirements. As the system requires analyzing the understanding level and the benefits student have got from a certain technique, there is a constant need for active student participation in the system. If the student does not attend the lecture sessions, submit assignments, perform practical sessions or answer quizzes, the teacher will not be able to comprehend how much the student has understood in a particular topic. Hence with active participation, the evaluation and further customization of the teaching–learning methodology becomes easy. ICT tools to a greater extent help in fulfilling this requirement of active student participation as they help students access a section and go through it along with taking up courses, projects, etc., and hence help in the implementation of the OBE framework.

7.6 CHALLENGES FACED BY TEACHERS AND STUDENTS IN THE EXISTING SYSTEM

To understand the challenges faced by the users (teachers/students) in the currently available ICT tools, a survey was created by designing a suitable questionnaire. The questionnaire was designed separately for teachers and students by considering various aspects involving the usage of ICT tools. From the survey, the difficulties faced by student and teacher community while using the available ICT tools were identified. Though the attempt was made to collect data pan India but the responses that were received were majorly from the western part of the country. On the backdrop of the pandemic that has spread across the world in the year 2020, there is an upsurge

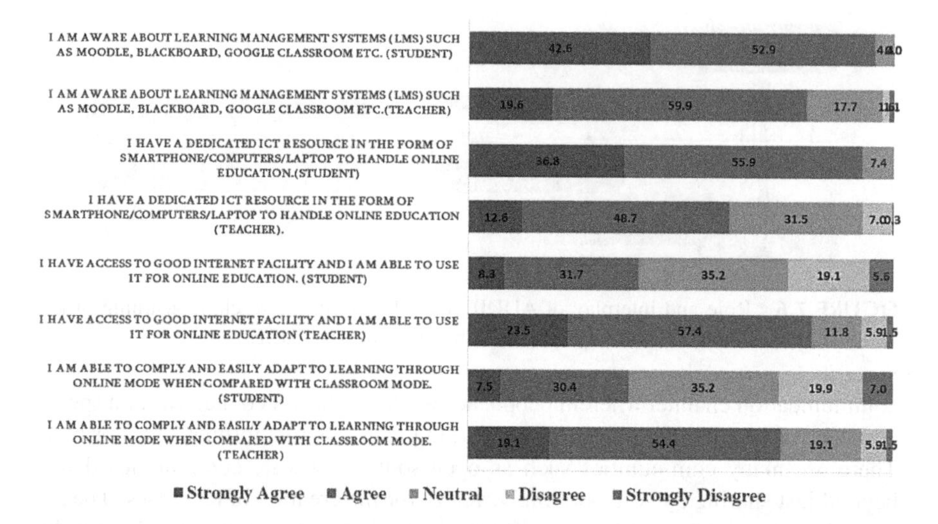

FIGURE 7.7 Survey analysis to check awareness, resource availability, internet access and adaptability of students and teachers for use of ICT in teaching learning process.

of usage in ICT tools which might have led to more awareness concerning the same. The authors feel that this is one of the important points to consider while reading the data recorded. The following were the major observations made from the survey conducted:

- As shown in Figure 7.7, a total of 38% of the students agreed that they are able to comply and easily adapt to learning through online mode when compared with classroom mode. On the other hand, a total of 73.5% of the teachers reported the ease of adapting to teaching through the incorporation of technology. 40% of the students and 81% of the teachers have responded to having access to a good internet facility and the ability to use the same for online education. 61% of the students and 93% of the teachers have responded to have a dedicated ICT resource in the form of smartphone/computers/laptop to handle online education. 80% of the students and 96% of the teachers have responded to being aware of LMS such as Moodle, Blackboard, Google Classroom, etc.
- As shown in Figure 7.8, a total of 71% of the students could easily access and use the course content, particularly, notes, videos, assignments and quiz through LMS for learning in engineering while 81% of the teachers reported to easily design and implement the same with the help of a LMS. However, only 21% of students and 40% of teachers reported engagement through discussion forums and using a peer assessment platform for aiding the teaching–learning process.
- As shown in Figure 7.9, the responses on the difficulties faced by students and teachers while using LMSs for learning activities exhibit that around 74% of students face the largest issue identified as a lack of good internet connection.

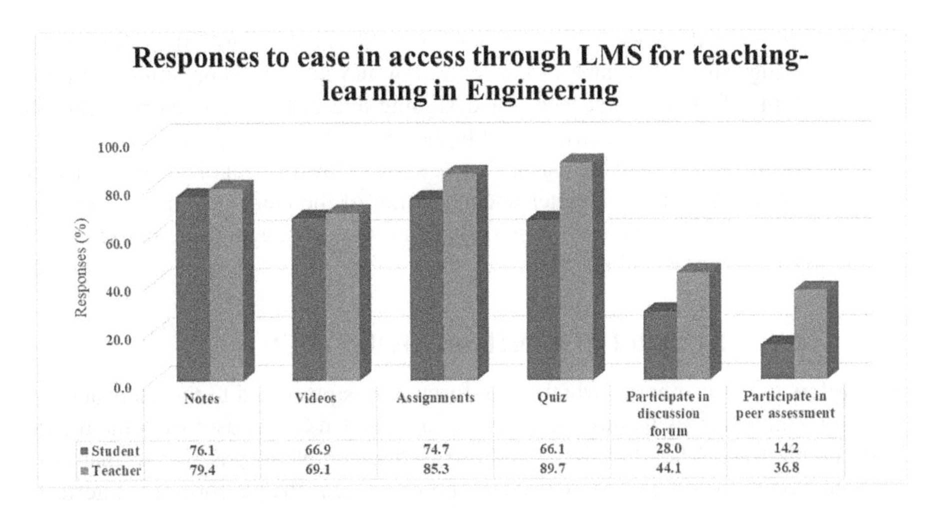

FIGURE 7.8 Survey analysis to ascertain the ease in access of selective course content through LMS for teaching–learning in engineering education.

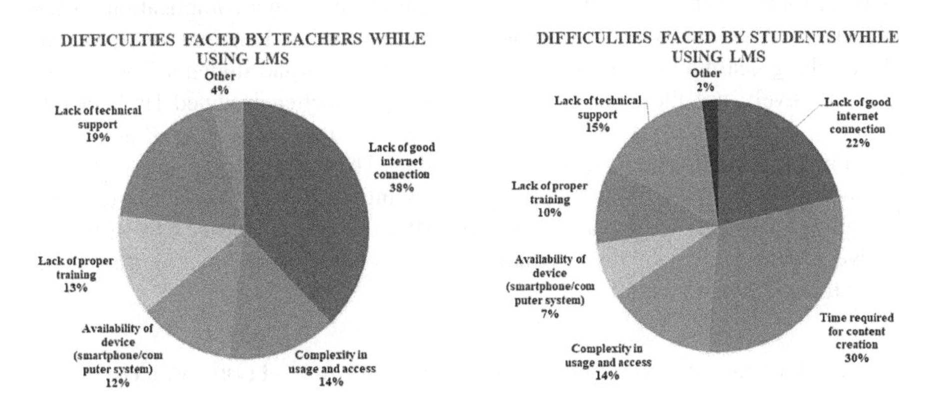

FIGURE 7.9 Survey analysis to understand the difficulties faced by teachers and students while using LMS for teaching–learning in engineering education.

On the other hand, the biggest problem that teachers face is the time required for the creation of good quality teaching–learning content suitable to be hosted on LMS platforms or for use in online teaching. Other issues including the complexity of usage and excess, availability of devices, lack of proper training and lack of technical support accounts to be an average of around 29% and 24% for students and teachers, respectively.

It is pertinent to mention that internet availability is pivotal in teaching and learning with the use of technology. There will be more clarity of the users over the issues mentioned above with good internet connectivity. The students have also reported other issues like transitioning difficulties from conventional classroom teaching to online teaching methods especially for mathematical and drawing content (prominent in engineering education), lack of concentration and non-appealing study environment, lack of efforts from teachers for designing and delivering content, technical issues like availability of electricity and higher live screen time. The survey reveals reporting of the issues from the teacher's side such as low turnout of students, lack of face-to-face interaction for a better understanding of the content transfer, interruptions due to sound disturbances on student's end, monitoring students presence and problems to teach practical courses.

7.7 HUMAN–COMPUTER INTERACTIONS (HCI)

HCI refers to the manner in which the information is presented to the users and the ways in which various features and services are availed by the user from the digital learning platform. The objective is to keep HCI as simple as possible so that even without having the domain knowledge on how to access the platform, i.e. interface, the user is able to easily understand, interpret and adaptively use the platform. It considers user experience to make the design user friendly which in turn can make the platform accessible with minimum efforts. A better interface would simply make it easier for the user to use and handle, thereby resolving major complications. These objectives of HCI have been incorporated in the education domain as well, to make the teaching–learning experience better for both teachers and students. The students can effectively use the content provided through a better designed HCI, thereby enabling them to easily keep a track of the syllabus they have covered along with providing information about other relevant topics. The forum offers a single platform for understanding views over a topic and discussing it from all possible angles. Even though it takes time to prepare the content, HCI provides an easier way of content delivery, design and availability making it more worthwhile by treating it as a template for teaching methodology.

7.7.1 Technology Acceptance Among the Teaching–Learning Community

The use of ICT in engineering education witnesses the increased usage of computers, laptops and smartphones, commonly for online lectures, submitting assignments/quiz, taking experiments virtually, or appearing for online examinations. This technological advancement has made us all witness the widest applicability of the technology in the domain of engineering education. With respect to the same, it becomes extremely important to impart knowledge about the use of technology, specifically computers in the engineering aspirants, right from the time they enter any domain of engineering program. Computer literacy is one of the key features that enable the use of technology for learning activities in an efficient manner as it has a deep impact (Mitra et al., 2000). The stakeholders involved in this are moving in the right direction of using it in a smarter way by exploring all the available options. To attain the enhancement in productivity, technology escorts the core changes which can be integrated. It permeates the learning environment with all the features in both teaching and learning such as involvement of digital classroom, teaching materials, effective assessment methods and continuous perpetual reinforcement. Also, technology helps

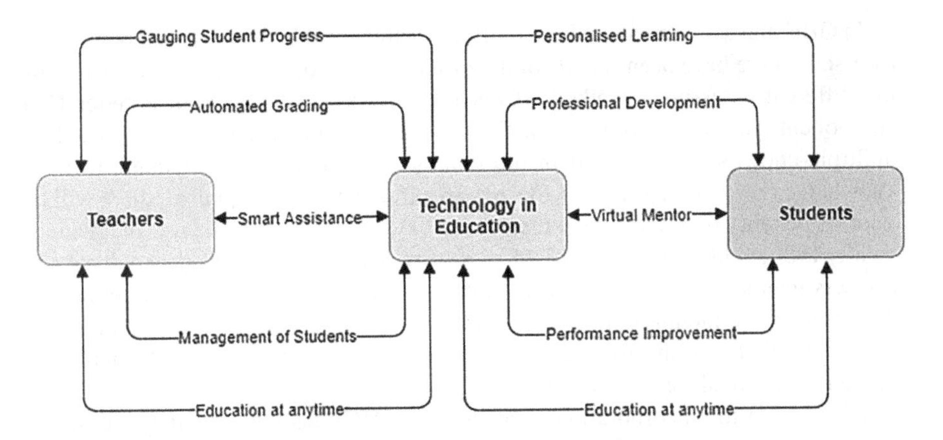

FIGURE 7.10 Features offered by the technology for students and teachers.

in getting the commitment of the students and expands learning opportunities to understand the content in a more effective way. It also has the capability to transmute teaching by assisting a new way of content delivery. A study was done to inspect the attitude toward educational technologies and both students and teachers have majorly been found to have a positive attitude toward acceptance of digitalization in the field of education (Alkan & Erdem, 2010). One of the similar reports suggests that 87.3% of the students in a survey liked to use technology and believed it could be effective in learning (Li, 2007). This reinforces the concept of empowerment of effective learning and advanced thinking by the use of technology in education. Figure 7.10 shows a flow chart of the methodology that is being adopted and used which incorporates technological advancements for the betterment of OBE. The management of student information, evaluation, grading and collecting feedback of students, etc., can all be performed by the use of technology in OBE and the teachers are hence relieved of the tiresome work which in turn boosts their efficiency. Due to the analysis performed by the implementation of technology in OBE, the teacher is able to vividly understand the level of attention/support each student needs, and hence both the teacher and students can meaningfully participate in the teaching–learning process.

7.7.2 Role of HCI in Assisting OBE

Students seem to perform their best when they have a structured learning curriculum with multiple checkpoints to analyze their progress and performance. This is basically a procedure for the assessment of the students' learning capabilities. In OBE, the assessment process must be aligned with the learning outcomes which means it should support the learners in their progress and validate the achievement of the intended LOs at the end of the process. In the traditional ways of assessments, there were constraints such as the dependability of results on the teacher, student and the resource availability which casts doubts on the accuracy of the results with respect to a qualitative measure of performance.

In OBE, the assessment method is strongly influenced by learning outcomes to be assessed. There have been reports of the different types of learning outcomes requiring different assessment methods (Crespo et al., 2010). LMSs with a better HCI component can help to implement the proper assessment methods by providing multiple-choices questions/fill in the blanks which can enable students to have knowledge type learning. If the LMS has an attractive HCI component, there will be a lot of benefits that both the students and the teachers can make use of. A high-quality HCI could be any web or LMS software interface that helps the students and teachers interact with it. The driving features such as unambiguous instructions on the usage of the modules, appropriate flow, adequate amount of information, good sound, visual effects, an attractive interface design, etc., will overcome the hardships of the conventional forms of learning.

Along with the incorporation of all features, HCI has the potential to ease the assessment process from the teacher's perspective as it helps them manage all the students and keep the track of them without manually maintaining report cards. HCI will enable tracking parameters like time stamps that determine the delay in the response that is registered by a student depicting the confidence of the students in the particular question and the domain the question was from. Also, features like handwriting recognition too could be added to the LMSs that would help the students who do not have good typing skills, facial expression recognition that could help to understand the behavior of the student and the teacher, voice and speech-tracking features of HCI can help understand the emotional state of the student and the teacher dealing with the LMS, thereby facilitating rational decision-making (Dzandu & Tang, 2015).

The generation of performance analytics such as comparison graphs also can be implemented to observe the comparative progress of a student with others for a certain type of assessment. The opportunities provided through ICT application methods are more inclusive and customized. The tools that ICT offers can help the students to appear for an evaluation/assignment on multiple occasions with immediate feedback. Also, there are no constraints involved regarding the location for appearing for an evaluation/assignment or the time to appear for the same, and malpractice issues while conduction of exam can be minimized using these tools. The dialogic engagement with students' progress can magnify teaching and assessment practices in support of all students' learning (Williams et al., 2014). To improve the students' knowledge, it is important to analyze how they derived a result or how they concluded or got a solution rather than simply focusing on the solution or conclusion. Hence, there is a need to analyze how well the student has understood a concept rather than analyzing how correct is the result, the student has derived. These facets need to be assessed for every student by incorporating the perfect assessment scheme. HCI plays a key role here as access to this perfect assessment scheme can be provided to the students. The conventional questionnaire, assignments, form filling or viva voce patterns of evaluating students that required manual evaluation would depend on a lot of human factors that may have a compromised accuracy due to human intervention and bias. The evaluation process becomes lesser cumbersome due to the effectiveness of HCI as it runs in a particular manner following a certain procedure or algorithm fed into it (Gawande, 2009).

7.7.3 Intelligent Tutor Systems and Latest Developments for Assisting OBE

The terminology of smart tutoring is increasingly gaining importance as a smart tutor helps the student in personalized learning leading to betterment in academic performance. It helps students to take control and become independent learners with better confidence and self-esteem. It also helps to finalize the most suitable content for the learner and helps in conducting the student profile-based exam by gathering data during the process of learning and can set the difficulty level of the questions to be included in the next test based on previous exam assessment results. It also supplements the learning process of each student, individually, in a virtual environment, as an all-time available guide (Rodrigues et al., 2010). AI reforms the culture for detailed reasoning and adaptive systems which help in managing and amalgamating content along with the separate evaluation of tutors (Sijing & Lan, 2018). To achieve improved teaching and corresponding assessment, AI is an essential key that is capable to help overcome the barriers prevailing in the current education system.

Smart assistance, a feature of ICT which includes the concept of AI, can easily track the student behavior toward the submission and can analytically help us know the level of understanding of each individual student. This may lead to proper counseling of the students to improve the performance by studying the facts from the assessment. It is rapidly growing and impacting the profound nature of services within education. The opportunities for AI in the education domain are expansive, scaling from customized learning plans and assistance provided by AI, automation of administrative tasks to providing uniform rights for all students. Also, it has the ability to push the process of acquiring education goals further ahead. AI plays a very crucial role from the teacher's point of view in checking the originality of the content and making students work on their own. These technologies can serve as the backbone to provide customized, adaptive and intelligent services for both teachers and students. Figure 7.11 gives a brief about the various ways AI can help in enhancing the features provided by ICT platforms. The use of various facilities such as a global classroom widens the scope for the development of a student. There is an automated grading system that assesses the student with the provision of help in identifying the level of student understanding. Also, the constructive and constant feedbacks taken from students help the teacher improve or modify the pattern of teaching as per the requirements.

7.7.4 AI's Integration in Teaching–Learning Processes and Assessment

To effectively deliver and understand the content, extra efforts are required by the teachers and students, respectively. Also, there are certain issues that make it difficult to achieve the desired goals which include revision of the topic, assigning the quiz and practical implementation. AI can play an effective role in teaching, learning and assessment processes. These aspects can be easier to a certain extent with the assistance provided by AI. In the flow of the learning experience, the features of AI-based applications not only meet the different learning requirements of students but also put the platform to absorb the content easily (Liang, 2020).

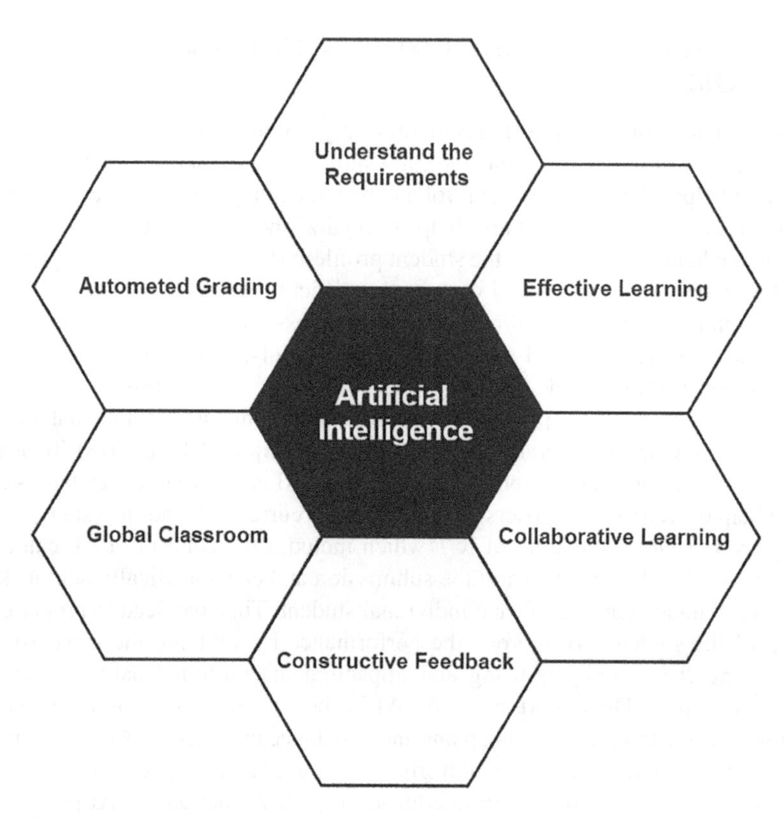

FIGURE 7.11 Beehive showing the various ways in which Artificial Intelligence complements ICT for engineering education.

In the conventional LMS, all the course content taught is made available for remote access to the student and the conduction of their exams along with the grades, i.e. reflection of student performance is made available by the teacher through the LMS. However, based on the accreditation requirements of OBE in engineering education, AI-based LMSs can be termed as the future of tomorrow. In an AI-based LMS, when the student is referring to the uploaded course content made available by the teacher, the time that students spend after a particular module (part) is taken into consideration to suggest the student's efficiency in that module/part or student's interest in the topic/s involved. In addition to the same, if there are assignments or evaluations as a part of that particular module, analysis of overall score can be made available which may comprise of the topics in which the student scored well in that module and the type of questions, the students did well in (Chang & Lu, 2019). From such a pattern, it can then be verified which elements of course content proved to be efficient in delivering the topic content and the methodology involved. On the basis of the analysis done using AI, customizations can be made in the course content offered by the teacher to the student in terms of assignment, quiz, videos and other associated study materials. The same pattern identification using AI could also suggest the courses in which student actually scores well or spends the most time learning, this engagement implying toward the interest of the student or difficulty in understanding. This analysis in manual form is difficult for teachers to perform individually for each student, whereas the AI-based LMS does it easily making use of the techniques that suggest patterns on the basis of prior knowledge, data and current statistics of the student thereby facilitating decision-making (Niu et al., 2019). Yet another benefit that the AI-based LMS platforms provide is that the student can avail the content in the preferred language. This would make the learning a lot easier compared to a mainstream language being used due to an easier understanding of the content.

It is difficult for teachers to evaluate and boost student performance without its detailed assessment. The teachers would be able to figure out an alternative way to cope up with the level of students by thorough analysis. The smart assessment system can help to make a rational assessment of the student learning process with the help of automatic grading, scientific thinking, personalized learning, scientific methods, etc. (Liang, 2020). The engineering education domain can better graduate performance in industries by availing the features of AI, and it will have an integral place in all services in the educational domain.

7.8 CONCLUSION

As the use of ICT in engineering education continues to grow, there is an increased acceptance of it among the teaching–learning community. This has led to improvement in various aspects of teaching, learning, administration, content creation, management and delivery. OBE has fast become the need in the hour in engineering education as it focuses on student-centric approach which is essential to impart and assess the skills acquired by a student as a part of the process that makes a good engineer. However, the implementation of OBE is demanding in nature as the involved activities are very difficult with only manual involvement. These activities when supplemented with the use of ICT make it easy for OBE implementation with the help of higher analytics. This, in turn, will also aid in tapping into the knowledge involved at higher levels of cognition thereby making it one of the only ways, it could be made possible. The use of advancements in technology can empower and enhance overall teaching–learning opportunities and the use of smart tools such as AI-based LMS shall be treated as a support system that allows taking individualized learning to the next level. The immediate challenges of increasing the level of education, universal access to all students and global learning with respect to the OBE standards can be met with the assistance of ICT in engineering education. Furthermore, the implementation of such an ICT-based system will alter the way teaching community will interact with the information which may change the role and responsibilities of the teaching community. The most important flavor of learning in engineering through trial and error will become more approachable and will aid in creating additional support for the student. In a nutshell, the future of education will see the theoretical frameworks such as instructional design and OBE work hand in hand with ICT for fulfilling the societal demands from engineering education.

REFERENCES

Abramova, N., & Grishchenko, N. (2020). ICTs, labour productivity and employment: Sustainability in industries in Russia. *Procedia Manufacturing*, *43*, 299–305. https://doi.org/10.1016/j.promfg.2020.02.161

Al-Ajlan, A. S. (2012). A comparative study between e-learning features. *Methodologies, Tools and New Developments for e-Learning*, 191–214.

Alkan, F., & Erdem, E. (2010). The attitudes of student teachers towards educational technologies according to their status of receiving teaching application lessons. *Procedia—Social and Behavioral Sciences*, *2*(2), 2523–2527. https://doi.org/10.1016/j.sbspro.2010.03.366

Andrich, D. (2002). A framework relating outcomes based education and the taxonomy of educational objectives. *Studies in Educational Evaluation*, *28*(1), 35–59. https://doi.org/10.1016/S0191-491X(02)00011-1

Asongu, S. A., & Odhiambo, N. M. (2020). Inequality and gender inclusion: Minimum ICT policy thresholds for promoting female employment in Sub-Saharan Africa. *Telecommunications Policy*, *44*(4), 101900. https://doi.org/10.1016/j.telpol.2019.101900

Barge, P., & Londhe, B. R. (2014). From teaching, learning to assessment: MOODLE experience at B'School in India. *Procedia Economics and Finance*, *11*(14), 857–865.

Barry Kissane, S. W. (1995). *Outcome-based Education: A review of the literature*. 55.

Bhattacharjee, B., & Deb, K. (2016). Role of ICT in 21st century's teacher education. *International Journal of Education and Information Studies*, *6*(1), 6.

Cavus, N., & Zabadi, T. (2014). A comparison of open source learning management systems. *Procedia-Social and Behavioral Sciences*, *143*, 521–526.

Chai, C. S., Ng, E. M., Li, W., Hong, H.-Y., & Koh, J. H. L. (2013). Validating and modelling technological pedagogical content knowledge framework among Asian preservice teachers. *Australasian Journal of Educational Technology*, *29*(1). https://doi.org/10.14742/ajet.174

Chang, J., & Lu, X. (2019). The study on students' participation in personalized learning under the background of artificial intelligence. *2019 10th International Conference on Information Technology in Medicine and Education (ITME)*, 555–558. https://doi.org/10.1109/ITME.2019.00131

Chootongchai, S., & Songkram, N. (2018). Design and development of SECI and Moodle online learning systems to enhance thinking and innovation skills for higher education learners. *International Journal of Emerging Technologies in Learning (IJET)*, *13*(03), 154–172.

Cox, K., Imrie, B. W., & Miller, A. (2014). *Student assessment in higher education: A handbook for assessing performance*. Routledge.

Crespo, R. M., Najjar, J., Derntl, M., Leony, D., Neumann, S., Oberhuemer, P., Totschnig, M., Simon, B., Gutiérrez, I., & Kloos, C. D. (2010). Aligning assessment with learning outcomes in outcome-based education. *IEEE EDUCON 2010 Conference*, 1239–1246. https://doi.org/10.1109/EDUCON.2010.5492385

Cristillo, L., Iter, N., & Assali, A. (2016). Sustainable leadership: Impact of an innovative leadership development program for school principals in Palestine. *American Journal of Educational Research*, *4*(2A), 37–42. https://doi.org/10.12691/education-4-2A-6

Cui, X., & Wang, Y. (2017). Current learning situation and future development of the credit-system-driven MOOC in higher education. *Digital Education*, *1*, 9.

Daher, W. (2012). Student teachers' perceptions of democracy in the mathematics classroom: Freedom, equality and dialogue. *Pythagoras*, *33*(2), 11. https://doi.org/10.4102/pythagoras.v33i2.158

David, R., & Krathwohl, L. W. A. (2001). *A taxonomy for learning, teaching, And assessing*. Addison Wesley.

Del Gaudio, B. L., Porzio, C., Sampagnaro, G., & Verdoliva, V. (2020). How do mobile, internet and ICT diffusion affect the banking industry? An empirical analysis. *European Management Journal*. https://doi.org/10.1016/j.emj.2020.07.003

Dewey, J., McMurray, C. A., & National Society for the Scientific Study of Education. (1904). *The relation of theory to practice in the education of teachers*. University of Chicago. http://archive.org/details/relationoftheory00dewe

Dzandu, M. D., & Tang, Y. (2015). Beneath a learning management system—Understanding the human information interaction in information systems. *Procedia Manufacturing*, *3*, 1946–1952. https://doi.org/10.1016/j.promfg.2015.07.239

Europeyou Association. (2020, November 19). What is Information and Communication Technology? *What is Information and Communication Technology?* http://europeyou.eu/es/what-is-information-and-communication-technology/

Florian, B., Glahn, C., Drachsler, H., Specht, M., & Fabregat Gesa, R. (2011). Activity-based learner-models for learner monitoring and recommendations in moodle. In C. D. Kloos, D. Gillet, R. M. Crespo García, F. Wild, & M. Wolpers (Eds.), *Towards Ubiquitous Learning* (111–124). Springer. https://doi.org/10.1007/978-3-642-23985-4_10

Gagné, R. M., Briggs, L. J., & Wager, W. W. (1992). *Principles of instructional design*. Harcourt Brace Jovanovich College Publishers. http://catalog.hathitrust.org/api/volumes/oclc/24219317.html

Gawande, V. (2009). Effective use of HCI in e-learning. *ResearchGate*, 17–18. https://www.researchgate.net/publication/228960510_Effective_Use_of_HCI_in_e-Learning

Gutiérrez, E., Trenas, M. A., Ramos, J., Corbera, F., & Romero, S. (2010). A new moodle module supporting automatic verification of VHDL-based assignments. *Computers & Education*, *54*(2), 562–577. https://doi.org/10.1016/j.compedu.2009.09.006

Haque, M. (2017). Outcome-based medical education – A brief commentary. *National Journal of Physiology, Pharmacy and Pharmacology*, *7*(9), 881–890.

Harden, R. M. (1999). AMEE Guide No. 14: Outcome-based education: Part 1-An introduction to outcome-based education. *Medical Teacher*, *21*(1), 7–14. https://doi.org/10.1080/01421599979969

Hopson, M. H., Simms, R. L., & Knezek, G. A. (2001). Using a technology-enriched environment to improve higher-order thinking skills. *Journal of Research on Technology in Education*, *34*(2), 109–119. https://doi.org/10.1080/15391523.2001.10782338

Hu, X., Hou, X., Lei, C.-U., Yang, C., & Ng, J. (2017). An outcome-based dashboard for moodle and Open edX. *Proceedings of the Seventh International Learning Analytics & Knowledge Conference*, 604–605. https://doi.org/10.1145/3027385.3029483

Huth, M., Vishik, C., & Masucci, R. (2017). 8—From risk management to risk engineering: challenges in future ICT systems. In E. Griffor (Ed.), *Handbook of System Safety and Security* (pp. 131–174). Syngress. https://doi.org/10.1016/B978-0-12-803773-7.00008-5

Iglesias-Pradas, S., Ruiz-de-Azcárate, C., & Agudo-Peregrina, Á. F. (2015). Assessing the suitability of student interactions from Moodle data logs as predictors of cross-curricular competencies. *Computers in Human Behavior*, *47*, 81–89. https://doi.org/10.1016/j.chb.2014.09.065

International Engineering Alliance. (2020, November 19). *Washington Accord*. https://www.ieagreements.org/accords/washington/

Jain, R., & Agrawal, R. (2007). ICT education and training services: Current scenario and future prospects in India. *Vision*, *11*(2), 41–55. https://doi.org/10.1177/097226290701100206

Jian-Feng, S., Yong-Gang, D., Xing-Dong, L. I., & Heng, Z. H. U. (2017). New teaching form for the course of engineering drawing based on OBE mode and Chinese MOOC platform-XuetangX. *DEStech Transactions on Social Science, Education and Human Science*, *icss*. 2017 4th International Conference on Social Science (ICSS 2017) ISBN: 978-1-60595-525-4 DOI: 10.12783/dtssehs/icss2017/19394. 78–82.

John, P. D., & Sutherland, R. (2004). Teaching and learning with ICT: New technology, new pedagogy? *Education, Communication & Information*, *4*(1), 101–107. https://doi.org/10.1080/1463631042000210971

Kastberg, S. E. (2003). Using bloom's taxonomy as a framework for classroom assessment. *The Mathematics Teacher*, *96*(6), 402.

Kerr, A. (2011). *Teaching and learning in large classes at Ontario universities: An exploratory study*. Higher Education Quality Council of Ontario.

Killen, D. R. (2000). *Outcomes-based education: Principles and possibilities*. Unpublished manuscript, University of Newcastle, Faculty of Education 1–24.

Klimov, B. F. (2012). ICT versus traditional approaches to teaching. *Procedia—Social and Behavioral Sciences*, *47*, 196–200. https://doi.org/10.1016/j.sbspro.2012.06.638

Koh, J. H. L., Chai, C. S., Benjamin, W., & Hong, H.-Y. (2015). Technological pedagogical content knowledge (TPACK) and design thinking: A framework to support ICT lesson design for 21st century learning. *The Asia-Pacific Education Researcher*, *24*(3), 535–543. https://doi.org/10.1007/s40299-015-0237-2

Kratochwill, T. R., Cook, J. L., Travers, J. F., & Elliott, S. N. (2000). *Educational psychology: Effective teaching, effective learning. Educational Psychology: Effective Teaching, Effective Learning* 3rd Edition. McGraw-Hill higher education, United States of America.

Kulkarni, T. P., Toksha, B. G., Bhosle, S. P., & Deshmukh, B. (2019). Analysing the impact of MOODLE and its modules on students learning, a case study in mechanical engineering. *Journal of Engineering Education Transformations*, *32*(3), 5661. https://doi.org/10.16920/jeet/2019/v32i3/143016

Lange, S., Pohl, J., & Santarius, T. (2020). Digitalization and energy consumption. Does ICT reduce energy demand? *Ecological Economics*, *176*, 106760. https://doi.org/10.1016/j.ecolecon.2020.106760

Lewis, A., & Smith, D. (1993). Defining higher order thinking. *Theory Into Practice*, *32*(3), 131–137. https://doi.org/10.1080/00405849309543588

Li, Q. (2007). Student and teacher views about technology: A tale of two cities? *Journal of Research on Technology in Education*, *39*(4), 377–397. https://doi.org/10.1080/15391523.2007.10782488

Liang, W. (2020). Development trend and thinking of artificial intelligence in education. *2020 International Wireless Communications and Mobile Computing (IWCMC)*, 886–890. https://doi.org/10.1109/IWCMC48107.2020.9148078

Majumdar, S. (2006). *Emerging trends in ICT for education & training*. Sense Publications. 13.

Malik, S., Rohendi, D., & Widiaty, I. (2019). *Technological pedagogical content knowledge (TPACK) with information and communication technology (ICT) integration: A literature review*. 498–503. https://doi.org/10.2991/ictvet-18.2019.114

McAlpine, M., Clark, R., Education, C. T., & Centre, C. (2002). *Principles of assessment*. CAA Centre, University of Luton.

McCahan, S., & Romkey, L. (2012). *Update on the University of Toronto graduate attribute process. Proceedings of the Canadian Engineering Education Association (CEEA)*.

McGurr, M. (2008). *Improving the flow of materials in a Cataloging Department: Using ADDIE for a project in the Ohio State University Libraries*. ResearchGate. https://www.researchgate.net/publication/290515732_Improving_the_flow_of_materials_in_a_Cataloging_Department_Using_ADDIE_for_a_project_in_the_Ohio_State_University_Libraries

Mecwan, A. I., Shah, D. G., & Fataniya, B. D. (2015). *Innovations in evaluation: An integral part of outcome based education. 2015 5th Nirma University International Conference on Engineering (NUiCONE)*, 1–5. https://doi.org/10.1109/NUICONE.2015.7449604

Merrill, M. D. (2002). First principles of instruction. *Educational Technology Research and Development*, *50*(3), 43–59. https://doi.org/10.1007/BF02505024

Mitra, A., Lenzmeier, S., Steffensmeier, T., Avon, R., Qu, N., & Hazen, M. (2000). Gender and computer use in an academic institution: Report from a longitudinal study. *Journal of Educational Computing Research*, *23*(1), 67–84. https://doi.org/10.2190/BG2M-A5ER-KV7Y-N0J5

Mlambo, S., Rambe, P., & Schlebusch, L. (2020). Effects of Gauteng province's educators' ICT self-efficacy on their pedagogical use of ICTS in classrooms. *Heliyon*, *6*(4), e03730. https://doi.org/10.1016/j.heliyon.2020.e03730

Mutanga, P., Nezandonyi, J., & Bhukuvhani, C. (2018). Enhancing engineering education through technological pedagogical and content knowledge (TPACK): A case study. *International Journal of Education and Development Using Information and Communication Technology*, *14*(3), 38–49.

Nadiyah, R. S., & Faaizah, S. (2015). The development of online project based collaborative learning using ADDIE model. *Procedia—Social and Behavioral Sciences*, *195*, 1803–1812. https://doi.org/10.1016/j.sbspro.2015.06.392

Niu, K., Cheng, C., Gao, H., & Zhou, X. (2019). Suggestions on accelerating the implementation of artificial intelligence technology in university information system. *2019 14th International Conference on Computer Science & Education (ICCSE)*, 767–770. https://doi.org/10.1109/ICCSE.2019.8845378

Otukile-Mongwaketse, M. (2018). Teacher centered dominated approaches: Their implications for todays inclusive classrooms. *International Journal of Psychology and Counselling*, *10*(2), 11–21. https://doi.org/10.5897/IJPC2016.0393

Ozdilek, Z., & Robeck, E. (2009). Operational priorities of instructional designers analyzed within the steps of the Addie instructional design model. *Procedia—Social and Behavioral Sciences*, *1*(1), 2046–2050. https://doi.org/10.1016/j.sbspro.2009.01.359

Palmer, E. (2007). *Engaging academics with a simplified analysis of their multiple-choice question (MCQ) assessment results*. https://www.academia.edu/30149683/Engaging_academics_with_a_simplified_analysis_of_their_multiple_choice_question_MCQ_assessment_results

Paterson Davenport, L. A., Davey, P. G., & Ker, J. S. (2005). An outcome-based approach for teaching prudent antimicrobial prescribing to undergraduate medical students: Report of a working party of the British society for antimicrobial chemotherapy. *Journal of Antimicrobial Chemotherapy*, *56*(1), 196–203. https://doi.org/10.1093/jac/dki126

Preston, R., Gratani, M., Owens, K., Roche, P., Zimanyi, M., & Malau-Aduli, B. (2020). Exploring the impact of assessment on medical students' learning. *Assessment & Evaluation in Higher Education*, *45*(1), 109–124. https://doi.org/10.1080/02602938.2019.1614145

Ramesh, V. M., Rao, N. J., & Ramanathan, C. (2015). Implementation of an intelligent tutoring system using Moodle. *IEEE Frontiers in Education Conference (FIE)*, *2015*, 1–9.

Rayón, A., Guenaga, M., & Núñez, A. (2014). Supporting competency-assessment through a learning analytics approach using enriched rubrics. *Proceedings of the Second International Conference on Technological Ecosystems for Enhancing Multiculturality*, 291–298. https://doi.org/10.1145/2669711.2669913

Rezgui, K., Mhiri, H., & Ghédira, K. (2014). Extending Moodle functionalities with ontology-based competency management. *Procedia Computer Science*, *35*, 570–579. https://doi.org/10.1016/j.procs.2014.08.138

Rodrigues, J. J. P. C., João, P. F. N., & Vaidya, B. (2010). EduTutor: An intelligent tutor system for a learning management system. *International Journal of Distance Education Technologies*, *8*(4), 66–80. https://doi.org/10.4018/jdet.2010100105

Sandeen, C. (2013). Integrating MOOCS into traditional higher education: The emerging "MOOC 3.0" Era. *Change: The Magazine of Higher Learning*, *45*(6), 34–39. https://doi.org/10.1080/00091383.2013.842103

Schiliro, F., & Choo, K.-K. R. (2017). Chapter 5—The role of mobile devices in enhancing the policing system to improve efficiency and effectiveness: A practitioner's perspective. In M. H. Au & K.-K. R. Choo (Eds.), *Mobile security and privacy* (85–99). Syngress. https://doi.org/10.1016/B978-0-12-804629-6.00005-5

Seels, B., & Richey, R. (1994). *Instructional technology: The definition and domains of the field*. Association for Educational Communications and Technology.

Shah, D. (2018, December 27). Edx's 2018: Year In Review — Class Central. *Class Central's MOOC Report*. https://www.classcentral.com/report/edx-2018-review/

Shahmir, S., Hamidi, F., Bagherzadeh, Z., & Salimi, L. (2011). Role of ICT in the curriculum educational system. *Procedia Computer Science*, *3*, 623–626. https://doi.org/10.1016/j.procs.2010.12.104

Shuman, L. J., Besterfield-Sacre, M., & McGourty, J. (2005). The ABET "professional skills"—Can they be taught? Can they be assessed? *Journal of Engineering Education*, *94*(1), 41–55.

Sijing, L., & Lan, W. (2018). Artificial intelligence education ethical problems and solutions. *2018 13th International Conference on Computer Science & Education (ICCSE)*, 1–5. https://doi.org/10.1109/ICCSE.2018.8468773

Song, J., Dong, Y., & Wang, W. (2020). *Construction of the Online Open Courses of Mechanical Drawing in Chinese Colleges Based on Chinese XuetangX MOOC Platform. Proceedings of the 2020 8th International Conference on Information and Education Technology*, 1–5. https://doi.org/10.1145/3395245.3395249

Spady, W. G. (1994). *Outcome-based education: Critical issues and answers*. American Association of School Administrators.

Spady, W. G. (1997). *Paradigm lost: Reclaiming America's educational future*. American Association of School Administrators.

Spady, W. G., & Marshall, K. J. (1991). Beyond traditional outcome-based education. *Educational Leadership*, *7*.

Stadin, M., Nordin, M., Broström, A., Magnusson Hanson, L. L., Westerlund, H., & Fransson, E. I. (2021). Technostress operationalised as information and communication technology (ICT) demands among managers and other occupational groups – Results from the Swedish Longitudinal Occupational Survey of Health (SLOSH). *Computers in Human Behavior*, *114*, 106486. https://doi.org/10.1016/j.chb.2020.106486

Tabulawa, R. (2003). International aid agencies, learner-centred pedagogy and political democratisation: A critique. *Comparative Education*, *39*(1), 7–26. https://doi.org/10.1080/03050060302559

Trust, T., & Pektas, E. (2018). Using the ADDIE model and universal design for learning principles to develop an open online course for teacher professional development. *Journal of Digital Learning in Teacher Education*, *34*(4), 219–233. https://doi.org/10.1080/21532974.2018.1494521

Tuckman, B. W. (1988). *Testing for teachers*. Harcourt College Publications.

Wang, Q. (2008). A generic model for guiding the integration of ICT into teaching and learning. *Innovations in Education and Teaching International*, *45*(4), 411–419. https://doi.org/10.1080/14703290802377307

Williams, P., Wray, J., Farrall, H., & Aspland, J. (2014). Fit for purpose: Traditional assessment is failing undergraduates with learning difficulties. Might eAssessment help? *International Journal of Inclusive Education*, *18*(6), 614–625. https://doi.org/10.1080/13603116.2013.802029

Xing, Q.-Y. (2018). Application of ADDIE model in instructional design of structural mechanics course. *DEStech Transactions on Social Science, Education and Human Science*, *esem*. https://doi.org/10.12783/dtssehs/esem2018/23914

Yassine, S., Kadry, S., & Sicilia, M.-A. (2016). A framework for learning analytics in moodle for assessing course outcomes. *2016 IEEE Global Engineering Education Conference (EDUCON)*, 261–266. https://doi.org/10.1109/EDUCON.2016.7474563

8 Academic Workers' Behaviors Toward Scientific Crowdsourcing

A Systematic Literature Review

Regina Lenart-Gansiniec
Jagiellonian University, Poland

CONTENTS

8.1 INTRODUCTION

In recent years, governments in many countries have implemented higher education reforms. Their goal was to provide more freedom to universities, to improve the quality of higher education and to professionalize the work of academic workers. Academics are expected to include members of the public in research projects, to establish cooperation with scientists from other research centers and to increase the

DOI: 10.1201/9781003102298-8

quality of research (Westra et al., 2011). In this context, the scientific enterprise is built on a foundation of trust. Society trusts that scientific research results are an honest and accurate reflection of a researcher's work. Researchers equally trust that their colleagues have gathered data carefully, have used appropriate analytic and statistical techniques, have reported their results accurately, and have treated the work of other researchers with respect.

Research on scientific crowdsourcing is extensive and has provided important insights into potential of scientific crowdsourcing (Riesch & Potter, 2014), requirements of the crowdsourcing platform (Schlagwein & Daneshgar, 2014), barriers to crowdsourcing projects conducted by researchers (Burgess et al., 2017; Law et al., 2017), identification of the type of crowdsourcing suitable for research purposes (Sauermann et al., 2019), potential of the virtual community for crowdsourcing, ways of motivating the virtual community (Baruch et al., 2016), and the characteristics of the virtual community involved in the endeavors of researchers (Franzoni & Sauermann, 2014).

However, there are still significant gaps (Beck et al., 2020). In particular, it is widely argued that employees, especially academic workers, play a key role in influencing effectiveness of these practices (Beck et al., 2019; Hui & Gerber, 2015). It is important to involve academic teachers and to convince them of their potential and effectiveness of scientific crowdsourcing (Riesch & Potter, 2014) because the teachers are an important and key element in school, and therefore their professional attributes are a significant factor in the processes of school change (Avidov-Ungar & Magen-Nagar, 2014). OECD (2015) "Making Open Science a Reality" and "EU Open Science Agenda" should contribute to academic workers reaching for scientific crowdsourcing; some researchers believed that they could do cutting-edge research with crowdsourcing but at the cost of derision of their peers and little or no recognition (Shirk et al., 2012). Higher schools may be restricted in their ability to crowdsource because they do not know how to trigger academic workers in a way that will encourage them to engage in crowdsourcing. For this reason, gaining deeper insight into the factors that influence behaviors of crowdsourcing is of great relevance as it will provide a more coherent picture of the crowdsourcing in higher education.

While several different theories have been developed to explain scientific crowdsourcing, there is no consensus on which motivations are most salient. According to Beck et al. (2020) "further micro-level research is still needed to develop a more complete understanding of capabilities, attitudes, values, characteristics, and motivations around different OIS practices, as well as the interplay of those elements with each other and with contextual factors". However, this is the knowledge needed if higher schools are to pursue crowdsourcing and align employee behaviors with that crowdsourcing.

The aim of this chapter is to summarize and review the research that has been carried out on academic workers' behaviors related to scientific crowdsourcing. This study followed a systematic review process focused on summarization of knowledge (Gough et al., 2012), reviewing 57 empirical research on scientific crowdsourcing. Building on Law et al. (2017), I understand scientific crowdsourcing as efforts that engage large numbers of people over the Web to help collect and process data. Behaviors of crowdsourcing can be characterized as the intentional creation, introduction, and application of new ideas within a work role, group, or organization, in order to benefit from role performance, the group, or the organization (Janssen, 2000).

Using the premises of the Theory of Planned Behavior (TPB), the aim of this chapter is to identify academic workers' behavior of scientific crowdsourcing. TPB is a broadly used theory to understand and predict individual human behavior of all kinds. The TPB is based on the assumption that human beings take available information into account and implicitly or explicitly consider the implications of their actions (Ajzen & Fishbein, 2005). According to the TPB three basic determinants are the function of intention to use behavior: attitude toward behavior, subjective norm to perform behavior, and perceived behavioral control over behavior. Thus the best predictor of volitional behavior is the formulated intention of getting involved in behavior (Ajzen, 1991). The intention is related to being motivated to perform a certain behavior, namely to what extent a given person is ready to try to make a given action happen and how much effort this person is willing to put into implementation of a certain behavior. This means that the more one is intended to act in a certain way, the more likely it is that this behavior materializes (Ajzen, 1991). However, before we analyze behavior, we should take into account the variables preceding the intention of contribution (Ferreira & Farias, 2018).

8.2 THEORETICAL BACKGROUND

8.2.1 CROWDSOURCING

The increasing interest in crowdsourcing among researchers was initiated in 2006 by the editor of "Wired" magazine, Howe. Howe defined it to be "the act of a company or institution taking a function once performed by employees and outsourcing it to an undefined (and generally large) network of people in the form of an open call" (Howe, 2006). Crowdsourcing is to apply principles of open-source software development, a form of peer production, to other contexts of social life, a collective effort in attaining knowledge, material, or financial resources to, again, produce knowledge, materials, and values.

Crowdsourcing has gained on popularity in management sciences owing to its potential (Afuah & Tucci, 2013), among others: business process improvement, creating open innovations, building of competitive advantage, access to experience, innovativeness, information, crowd skills and work, which are located outside the organization (Aitamurto et al., 2011). It started to be linked to initiating collaboration and relations with virtual communities, further on using their wisdom to solve problems, participation management, increasing transparency and openness of public organizations (Brabham, 2008). Crowdsourcing also enables crisis management, expands existing activity and offer of the organization, creates the organization's image, improves communication with the surroundings, and optimizes costs of the organization's activity. It also enables access to knowledge and creativity resources and facilitates acquiring of new contents and data (Majchrzak & Malhotra, 2013).

8.2.2 Scientific Crowdsourcing

Scientific crowdsourcing has become an important part of the changing science landscape. Scientific crowdsourcing is based on online collaborative, massively collaborative science (Wiggins & Crowston, 2011), public participation (English et al., 2018), and volunteer engagement in knowledge creation (Franzoni & Sauermann, 2014). In this context, scientific crowdsourcing could be conceived as a set of workflow tasks, assets, processes, and outputs (Hedges & Dunn, 2018). Scientific crowdsourcing is a new way of contemporary scientific research activities, an example of opening of science and research, an alternative to research projects, a strategy for organizing the work of a researchers, and tool for research. Scientific crowdsourcing enables new collaborative forms of knowledge creation; scientific crowdsourcing is an online content creation tool (Doan et al., 2011; Kalev et al., 2013) communicating academic teachers with each other and with people from outside the scientific community (Scanlon et al., 2014), collecting or classifying data (Beck et al., 2019). It is also the practice of obtaining participants, services, ideas, or content by soliciting contributions from a large group of people, especially via the Internet (Brown & Allison, 2014). Scientific crowdsourcing facilitates the process of collecting, processing, and analyzing research data (Law et al., 2017); enlisting participants for surveys, research, experiments, panels, focus groups, statistical analyses, transcriptions (Schlagwein & Daneshgar, 2014); and generating innovative research questions, hypotheses, research proposals, testing research at an early stage. Scientific crowdsourcing also allows for reducing costs of conducting research, providing the researchers involved with funding, establishing cooperation and seeking collaborators for joint research (Crowston, 2012), obtaining assessment and opinions on the concept of a given research project or an article (Ipeirotis et al., 2010; Uhlmann et al., 2019), solving problems arising in the course of writing an article or conducting research (Hevner et al., 2013), determining the reliability and generalization of the results (Pan et al., 2017), and disseminating the results (Beck et al., 2019).

8.2.3 Techniques for Scientific Crowdsourcing

The possibility of including an unlimited number of participants in a research project and for researchers to interact with project team members is possible, thanks to an open invitation via a crowdsourcing platform. Crowdsourcing platforms act as an orchestrator and an intermediary between the researcher and members of the virtual community, and regulate, coordinate the provision and development of knowledge valuable for the researcher. Crowdsourcing platforms also allow you for inviting and keeping in touch with the virtual community. Generally speaking, they bring together the distributed workforce of the virtual community to gather resources, process information, or create new content.

There are many crowdsourcing platforms around the world that specialize in inviting virtual communities to a variety of research-related tasks, for example,

EpiCollect.net, CrowdCurio, GalaxyZoo, eBird. While there are many crowdsourcing platforms, Amazon's Mechanical Turk (MTurk) (http://www.mturk.com) is by far the most used social science research platform. MTurk is a universal platform—easy to use, with versatile use at all stages of the systematic literature review. MTurk ensures the recruitment of employee-participants to perform tasks called Human Intelligence Tasks (HIT). HITs are often short, repetitive, and small in scope.

Currently, over 750,000 people are registered on the MTurk platform and they take part in various projects or express their willingness to participate. About 80% of them live in the United States and are under 50, while 51% have a university degree. Researchers can also use this platform to recruit volunteers to complete surveys, participate in experiments, and conduct content analysis. One of the main reasons why MTurk is attractive to researchers is its ability to collect data quickly. For example, a survey that requires 300 respondents can be completed within hours. Moreover, the pool of MTurka's respondents is also much more diverse than the typical sample of students. Additionally, data collection costs are typically lower than those charged on other crowdsourcing platforms.

8.3 METHODOLOGY

8.3.1 Search Strategy

Creating and expanding new knowledge requires knowledge of its current state and identification of existing recommendations for further research. Systematic review differs from more traditional reviews because it uses systematic and explicit methods to identify, select, and critically appraise relevant research, and to collect and analyze data from the studies that are included in the review (Moher et al., 2009). I have chosen a systematic literature review because it is repeatable and clear, and it includes several clear steps. It also enables the identification and synthesis of the results of all major research and theoretical approaches, which in turn enables the identification of existing cognitive and research gaps (Sekaran & Bougie, 2010) (Tranfield et al., 2003). In addition, systematic literature reviews covering higher education (Tight, 2018) are becoming more and more popular. They are also becoming a requirement in business and management research (Fisch & Block, 2018). The literature emphasizes the need to conduct systematic literature reviews in the field of crowdsourcing in higher education (Lukyanenko et al., 2019).

Two strategies were used to identify eligible articles. Firstly, I carried out an electronic search in Scopus and Web of Science restricted to Social Science Citation Index in title, abstract, and keywords of articles. The term [scientific crowdsourcing*] was searched along with the term [behavior*]. This strategy resulted in a total of 551 articles. Second stage of the literature search involved searching through articles on crowdsourcing published in the following ten leading higher education journals: *Review of Educational Research*, *Review of Higher Education*, *Journal of Higher Education*, *Higher Education Quarterly*, *Higher Education Research &*

Development, Studies in Higher Education, Higher Education Policy, Higher Education, Research in Higher Education, Internet and Higher Education. When selecting them, they were guided by the indications (Tight, 2018) and results of Scimago Journal & Country Rank 2018. Same term strings were selected in title, abstract, keywords, and full text. This strategy does not produce studies.

8.3.2 Identification of Sources

I employed "Preferred Reporting Items for Systematic Review and Meta-Analyses" (PRISMA) (Moher et al., 2009) to identify eligible studies. The articles should meet the following six criteria to be included in the review:

1. Field: Articles should study crowdsourcing in research in higher education. I defined higher education as "all universities, colleges of technology and other institutions of post-secondary education, whatever their source of finance or legal status; all research institutes, experimental stations and clinics operating under the direct control of or administered by or associated with higher education institutions" (OECD, 2002).
2. Topic: The studies included need to provide details on the participant behaviors of crowdsourcing in higher education.
3. Topic: Only empirical studies were eligible for literature review (literature reviews and conceptual works were not included) because we are interested in empirical evidence concerning science crowdsourcing. All research projects were acceptable (e.g., questionnaire, case study, experiment), but case studies that were purely illustrative were excluded. We also excluded systematic reviews (Strang & Simmons, 2018) to avoid double-counting of research. Conference papers or proceedings, article reviews, book chapters, and unpublished theses and dissertations were not included.
4. Language: Only publications written in English were taken into account.
5. Year of publication: Articles that were published between 2006 and September 2020 were searched and included. The starting date includes the first appearance of the concept of crowdsourcing in literature (Howe, 2006).
6. Publication types: Only international, peer-reviewed, full-text publications were included.

The study selection methods and process are presented in the PRISMA flow diagram (Figure 8.1) as follows: Step 1 identified a total of 551 articles (Scopus: 470; Web of Science: 81). Step 2 reduced the sample by 449 articles because of redundancies between the different search engines. In Step 3, the abstracts of the remaining 325 articles were checked regarding the inclusion criteria, leading to the removal of a further 645 articles, which left 78 articles. Finally, studies included in the qualitative synthesis amounted to 57.

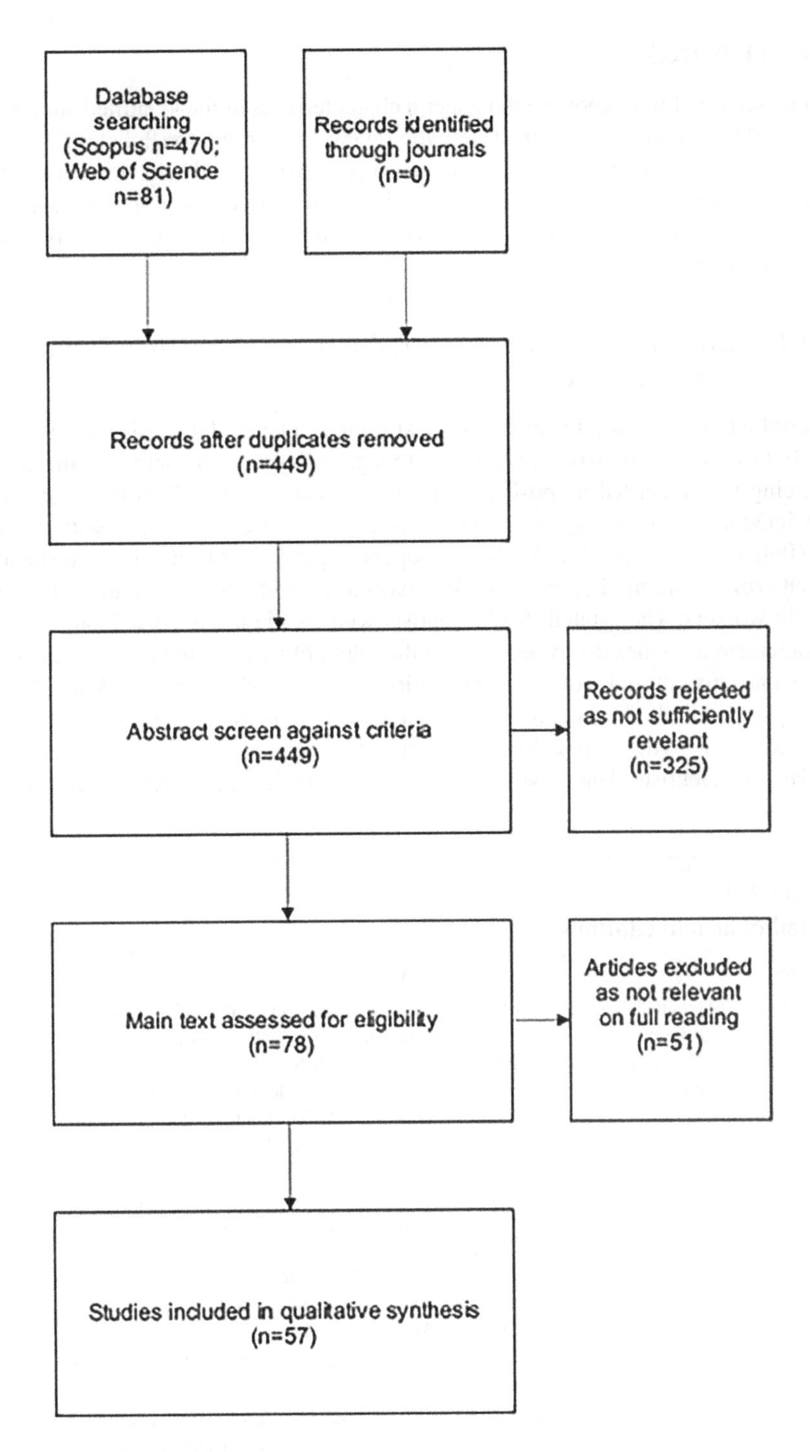

FIGURE 8.1 PRISMA flow diagram.

8.4 FINDINGS

In this section, I will report on the general characteristics of the identified studies and I present the findings from the systematic review. According to (Poliakoff & Webb, 2007), past behavior, attitude, subjective norms, and perceived behavioral control are important for crowdsourcing. Therefore, each section details a specific aspect of behavior of scientific workers, especially attitude, subjective norms, and perceived behavioral control (Table 8.1.)

8.4.1 Attitudes of Academic Workers' Behaviors Toward Scientific Crowdsourcing

According to Ajzen and Fishbein, 2005, attitude is a learned disposition to a consistently favorable or unfavorable response to a given object (whether scientific crowdsourcing was regarded as positive). In this context, scientists' attitudes are some predicted intentions to engage with crowdsourcing, especially activity with the public (Poliakoff & Webb, 2007). Mostly, papers suggested that scientists are skeptical about crowdsourcing. For example, Poliakoff and Webb (2007) examined 1000 scientific workers. They stated that "scientists who decide not to participate in public engagement activities do so because (a) they did not participate in the past, (b) they have a negative attitude toward participation, (c) they feel that they lack the skills to take part, and (d) they do not believe that their colleagues participate in public engagement activities" (p. 259). Other authors, Riesch and Potter (2014) interviewed 30 British scientists. The survey showed negative beliefs in the respondents toward

TABLE 8.1
Detail of article citations

Themes	Authors	No.
Attitude	(Alghamdi et al., 2015); (Aristeidou et al., 2017); Beck et al. (2020); Beck et al. (2019); (Behrend et al., 2011); (Bonney et al., 2015); (Brasseur et al., 2019); (Chen et al., 2011); (Cullina et al., 2014); (English et al., 2018); Franzoni, Sauermann (2014); (Füchslin et al., 2019); Kim, Adler (2015); (Lifshitz-Assaf, 2017); (Lind et al., 2017); (Mazumdar et al., 2017); (Michel et al., 2015); (Nov et al., 2010); (Nov et al., 2011); (Olson & Rosacker, 2013); (Owen et al., 2012); (Pan et al., 2017); (Parrick & Chapman, 2020); (Petersen, 2013); Riesch et al. (2013); (Rotman et al., 2012); (Sauermann & Franzoni, 2015); Sauermann et al. (2019); (Sauermann et al., 2020); (Savio et al., 2016); (Scheliga et al., 2016); (See et al., 2016); (Shank, 2015); (Sheehan, 2018); (Smith et al., 2015); (Steelman et al., 2014); (Stewart et al., 2017); (Stritch et al., 2017); (Uhlmann et al., 2019); (Wang & Yu, 2019); (Watson & Floridi, 2016); (Wechsler, 2014); (Wexler, 2011); (Wheat et al., 2013); (Williams, 2013)	44
Subjective norms	(Guinan et al., 2013); Hecker et al. (2018); Hui, Gerber (2015); Poliakoff, Webb (2007); (Shirk et al., 2012)	5
Perceived behavioral control	Burgess et al. (2017); (Can et al., 2017); (Cunha et al., 2017); Kim, Adler (2015); Law et al. (2017); Poliakoff, Webb (2007); (Schlagwein & Daneshgar, 2014); (Wiggins & Crowston, 2011)	8

the potential and effectiveness of crowdsourcing. In turn, Schlagwein and Daneshgar (2014) organized a series of focus studies with 28 researchers from Asia and the Pacific. Most of the respondents had negative views on the usefulness of crowdsourcing for research purposes. Kim and Adler (2015) found that scientific workers' motivations and norm of data sharing currently support social scientists' data sharing. In this context, scientists' data sharing behaviors are mainly driven by personal motivations (i.e., perceived career benefit and risk, perceived effort, and attitude toward data sharing) and perceived normative pressure. Funding agencies' pressure, journals' pressure, and availability of data repository were not found to be significant factors in influencing social scientists' data sharing. Law et al. (2017) identified some barriers in crowdsourcing projects implemented by the researchers. They indicate, among other things, lack of awareness among researchers about the value of crowdsourcing. Law et al. (2017) interviewed 18 researchers. Their research demonstrates that the reasons why crowdsourcing may be poorly adapted to the needs of researchers may include: feasibility, desirability, and utility. Beck et al. (2019) analyzed the crowdsourcing platform, on which they invited individuals with personal or professional experiential knowledge on accidental injuries to participate for a small monetary incentive. A total of 722 people were examined. Beck et al. (2019) suggested that "members tend to lack some of the knowledge that is required to formulate useful RQs, as reflected in a high share of ill-structured research questions".

8.4.2 Subjective Norms of Academic Workers' Behaviors Toward Scientific Crowdsourcing

Subjective norms are the individual's beliefs about whether significant others think he or she should engage in the behavior and are assumed to capture the extent of perceived social pressures exerted on individuals to engage in a certain behavior (Ajzen, 1991). Poliakoff and Webb (2007) suggested that "although scientists may have a positive attitude toward participating in public engagement activities and believe that significant others would approve of their taking part, they may not feel confident about their ability to engage with the public".

In this context, subjective norms are also important in scientific crowdsourcing. For, example, Shirk et al. (2012), based on narrative research with nine crowdsourcing researchers, show that some researchers believed that they could do cutting-edge research with crowdsourcing, but at the cost of derision of their peers and little or no recognition. According to Guinan et al. (2013), scientists funding decisions are the credibility of the investigator in the field and the potential impact and feasibility of the proposed research.

8.4.3 Perceived Behavioral Control of Academic Workers' Behaviors Toward Scientific Crowdsourcing

Perceived behavioral control is assumed to reflect past experience with the performance of the behavior and anticipated obstacles that could inhibit behavior (Ajzen, 1991). Schlagwein and Daneshgar (2014) suggested that it was prohibitively difficult for the majority of researchers to effectively use existing crowdsourcing platforms for performing research tasks. The problems refer to technical skills, programming that are required to critical functionalities for researchers that are missing, and/or to the crowd that is not suitable for research tasks. Another problem is connected with the lack of a crowdsourcing platform for research, which also prevents effective collaborative research by academics and collaborative research by non-academics. Some authors recommended Amazon Mechanical Turk as an academic research platform for data collection (Sheehan, 2018), exploratory analyses and survey construction, measurement refinement of latent constructs, experiments, and longitudinal research (Shank, 2015; Stritch et al., 2017). Others, Beck et al. (2020) and Poliakoff and Webb (2007) suggested that perceived behavioral control or individual scientists' perception seemed to predict intentions to engage scientific crowdsourcing more reliably than institutional resources, platform, training, or education support.

8.5 DISCUSSION AND FUTURE RESEARCH DIRECTIONS

The aim of this chapter is to summarize and review the research that has been carried out on academic workers' behaviors of scientific crowdsourcing. Based on TPB, the analyzed research can be categorized into three main research types: attitudes, subjective norms, and perceived behavioral control. The majority of the reviewed studies have focused on attitudes of academic workers toward scientific crowdsourcing. Mostly, papers suggested that scientists are skeptical or they demonstrate negative attitude toward scientific crowdsourcing, especially in the context of views on the usefulness and qualitative of crowdsourcing for research purposes. Research focusing on subjective norms shows that subjective norms are not important for academic workers, especially while getting engaged in scientific crowdsourcing. In this context, academic crowdsourcing is believed to allow for cutting-edge research with crowdsourcing, but at the cost of derision of their peers and little or no recognition—and therefore it is not taken up. Finally, in a study of a perceived behavioral control it was found out that the majority of researchers effectively used existing crowdsourcing platforms for performing research tasks. Even though platform or other tools might seem encouraging for users, perceived behavioral control can also have negative outcomes through making academic workers feel obstacles that could inhibit engagement in scientific crowdsourcing.

As with any emerging area of study and practice, there are a few research questions that require exploration. These include, for example, the following one: what behavioral factors determine academic teachers' recourse to scientific crowdsourcing? In addition, there is a need to better understand academic workers' behaviors of scientific crowdsourcing. These identified questions and research are not intended to be comprehensive, but rather to suggest future areas of research academic workers continue to explore the use and adoption of scientific crowdsourcing. More empirical research is needed to gain a deeper understanding about behavioral antecedents of scientific crowdsourcing. There is not enough identification research considering all types of scientific crowdsourcing, behavioral antecedents of scientific crowdsourcing from the perspective of academic teachers. Recognition of antecedents and approval of scientific crowdsourcing may help to better understand the potential of scientific crowdsourcing and its use.

ACKNOWLEDGMENT

This project was financed from the funds provided by the National Science Centre, Poland awarded on the basis of decision number DEC-2019/35/B/HS4/01446.

REFERENCES

Afuah, A., & Tucci, C. (2013). Value capture and crowdsourcing. *Academy of Management Review*, *38*, 457–460. https://doi.org/10.5465/amr.2012.0423

Aitamurto, T., Leiponen, A., & Tee, R. (2011). *The promise of idea crowdsourcing–benefits, contexts, limitations*. Nokia Ideas Project.

Ajzen, I. (1991). The theory of planned behavior. *Organizational Behavior and Human Decision Processes*, *50*, 179–211. https://doi.org/10.1016/0749-5978(91)90020-T

Ajzen, I., & Fishbein, M. (2005). The Influence of attitudes on behavior. In (Vol. 173, pp. 173–221).

Alghamdi, E., Aljohani, N., Alsaleh, A., Bedewi, W., & Basheri, M. (2015). CrowdyQ: A virtual crowdsourcing platform for question items development in higher education. *Proceedings of the 17th International Conference on Information Integration and Web-based Applications & Services*. https://doi.org/10.1145/2837185.2843852

Aristeidou, M., Scanlon, E., & Sharples, M. (2017). Profiles of engagement in online communities of citizen science participation. *Computers in Human Behavior*, *74*, 246–256. https://doi.org/10.1016/j.chb.2017.04.044

Avidov-Ungar, O., & Magen-Nagar, N. (2014). Teachers in a changing world: Attitudes toward organizational change. *Journal of Computers in Education*, *1*(4), 227–249. https://doi.org/10.1007/s40692-014-0014-x

Baruch, A., May, A., & Yu, D. (2016). The motivations, enablers and barriers for voluntary participation in an online crowdsourcing platform. *Computers in Human Behavior*, *64*, 923–931. https://doi.org/10.1016/j.chb.2016.07.039

Beck, S., Bergenholtz, C., Bogers, M., Brasseur, T.-M., Conradsen, M. L., Di Marco, D., Distel, A. P., Dobusch, L., Dörler, D., Effert, A., Fecher, B., Filiou, D., Frederiksen, L., Gillier, T., Grimpe, C., Gruber, M., Haeussler, C., Heigl, F., Hoisl, K., Hyslop, K., Kokshagina, O., LaFlamme, M., Lawson, C., Lifshitz-Assaf, H., Lukas, W., Nordberg, M., Norn, M. T., Poetz, M., Ponti, M., Pruschak, G., Pujol Priego, L., Radziwon, A., Rafner, J., Romanova, G., Ruser, A., Sauermann, H., Shah, S. K., Sherson, J. F., Suess-Reyes, J., Tucci, C. L., Tuertscher, P., Vedel, J. B., Velden, T., Verganti, R., Wareham, J., Wiggins, A., & Xu, S. M. (2020). The Open Innovation in Science research field: A collaborative conceptualisation approach. *Industry and Innovation*, 1–50. https://doi.org/10.1080/13662716.2020.1792274

Beck, S., Brasseur, T.-M., Poetz, M. K., & Sauermann, H. (2019). What's the problem? How crowdsourcing contributes to identifying scientific research questions. *Academy of Management Proceedings*, *2019*(1), 15282. https://doi.org/10.5465/ambpp.2019.115

Behrend, T. S., Sharek, D. J., Meade, A. W., & Wiebe, E. N. (2011). The viability of crowdsourcing for survey research. *Behavior Research Methods*, *43*(3), 800. https://doi.org/10.3758/s13428-011-0081-0

Bonney, R., Phillips, T. B., Ballard, H. L., & Enck, J. W. (2015). Can citizen science enhance public understanding of science? *Public Understanding of Science*, *25*(1), 2–16. https://doi.org/10.1177/0963662515607406

Brabham, D. (2008). Crowdsourcing as a model for problem solving: An introduction and cases. *Convergence: The International Journal of Research Into New Media Technologies*, *14*, 75–90. https://doi.org/10.1177/1354856507084420

Brasseur, T.-M., Beck, S. C., Sauermann, H., & Pötz, M. (2019). *Crowdsourcing research questions? Leveraging the crowd's experiential knowledge for problem finding. DRUID Academy Conference 2019 at Aalborg University*, Denmark January 16-18, 2019, https://conference.druid.dk/acc_papers/gucive3z7vfzso1kv6ydmzxawl8rqu.pdf

Brown, A. W., & Allison, D. B. (2014). Using crowdsourcing to evaluate published scientific literature: Methods and example. *PLOS ONE*, *9*(7), e100647. https://doi.org/10.1371/journal.pone.0100647

Burgess, H. K., DeBey, L. B., Froehlich, H. E., Schmidt, N., Theobald, E. J., Ettinger, A. K., HilleRisLambers, J., Tewksbury, J., & Parrish, J. K. (2017). The science of citizen science: Exploring barriers to use as a primary research tool. *Biological Conservation*, *208*, 113–120. https://doi.org/https://doi.org/10.1016/j.biocon.2016.05.014

Can, Ö., D'Cruze, N., Balaskas, M., & Macdonald, D. (2017). Scientific crowdsourcing in wildlife research and conservation: Tigers (Panthera tigris) as a case study. *PLOS Biology*, *15*, e2001001. https://doi.org/10.1371/journal.pbio.2001001

Chen, J., Menezes, N., & Bradley, A. (2011). Opportunities for crowdsourcing research on Amazon Mechanical Turk. *Interfaces*, *5*(3), 1.

Crowston, K. (2012). Amazon Mechanical Turk: A research tool for organizations and information systems scholars. In *Shaping the future of ict research. methods and approaches* (pp. 210–221). Springer. https://doi.org/10.1007/978-3-642-35142-6-14

Cullina, E., Morgan, L., & Conboy, K. (2014). *The development of a public/private model for the crowd- funding and crowdsourcing of scientific research projects. 3rd International workshop on IT Artefact Design & Workpractice Improvement*, 2 June, 2014, Friedrichshafen.

Cunha, D., Marques, J., De Resende, J., De Falco, P., De Souza, C., & Loiselle, S. (2017). Citizen science participation in research in the environmental sciences: Key factors related to projects' success and longevity. *Anais da Academia Brasileira de Ciências*, *89*. https://doi.org/10.1590/0001-3765201720160548

Doan, A., Ramakrishnan, R., & Halevy, A. Y. (2011). Crowdsourcing systems on the world-wide web. *Communication of the ACM*, *54*(4), 86–96. https://doi.org/10.1145/1924421.1924442

English, P. B., Richardson, M. J., & Garzón-Galvis, C. (2018). From crowdsourcing to extreme citizen science: Participatory research for environmental health. *Annual Review of Public Health*, *39*, 335–350. https://doi.org/10.1146/annurev-publhealth-040617-013702

Ferreira, G. d. D., & Farias, J. S. (2018). The motivation to participate in citizen-sourcing and hackathons in the public sector. *BAR—Brazilian Administration Review*, *15*, 1–22.

Fisch, C., & Block, J. (2018). Six tips for your (systematic) literature review in business and management research. *Management Review Quarterly*, *68*(2), 103–106.

Franzoni, C., & Sauermann, H. (2014). Crowd science: The organization of scientific research in open collaborative projects. *Research Policy*, *43*(1), 1–20.

Füchslin, T., Schäfer, M., & Metag, J. (2019). Who wants to be a citizen scientist? Identifying the potential of citizen science and target segments in Switzerland. *Public Understanding of Science*, *28*. https://doi.org/10.1177/0963662519852020

Gough, D., Thomas, J., & Oliver, S. (2012). Clarifying differences between review designs and methods. *Systematic reviews*, *1*, 28. https://doi.org/10.1186/2046-4053-1-28

Guinan, E., Boudreau, K., & Lakhani, K. (2013). Experiments in open innovation at Harvard Medical School. *MIT Sloan Management Review*, *54*, 45–52.

Hecker S, Haklay M, Bowser A, et al. (2018) Innovation in Open Science, Society and Policy – Setting the Agenda for Citizen Science. In: Hecker S, Haklay M, Bowser A, et al. (eds) *Citizen Science: Innovation in Open Science, Society and Policy*. London: UCL Press, 1–24.

Hedges, M., & Dunn, S. (2018). Chapter 1—Introduction: Academic crowdsourcing from the periphery to the centre. In M. Hedges & S. Dunn (Eds.), *Academic crowdsourcing in the humanities* (pp. 1–12). Chandos Publishing. https://doi.org/10.1016/B978-0-08-100941-3.00001-2

Hevner, A., March, S., Park, J., & Ram, S. (2013). *Design science in information systems research*. https://doi.org/10.2307/25148625

Howe, J. (2006). The rise of crowdsourcing. *Wired Magazine*, *14*, 1–4.

Hui, J. S., & Gerber, E. M. (2015). *Crowdfunding science: Sharing research with an extended audience*. *Proceedings of the 18th ACM Conference on Computer Supported Cooperative Work & Social Computing*, Vancouver, BC, Canada. https://doi.org/10.1145/2675133.2675188

Ipeirotis, P. G., Provost, F., & Wang, J. (2010). *Quality management on Amazon Mechanical Turk*. *Proceedings of the ACM SIGKDD Workshop on Human Computation*, Washington DC. https://doi.org/10.1145/1837885.1837906

Janssen, O. (2000). Job demands, perceptions of effort-reward fairness and innovative work behaviour. *Journal of Occupational and Organizational Psychology*, *73*(3), 287–302. https://doi.org/10.1348/096317900167038

Kalev, L., Shaowen, W., Guofeng, C., Anand, P., & Eric, S. (2013). Mapping the global Twitter heartbeat: The geography of Twitter. *First Monday*, *18*(5). https://doi.org/10.5210/fm.v18i5.4366

Kim, Y., & Adler, M. (2015). Social scientists' data sharing behaviors: Investigating the roles of individual motivations, institutional pressures, and data repositories. *International Journal of Information Management*, *35*(4), 408–418. https://doi.org/10.1016/j.ijinfomgt.2015.04.007

Law, E., Gajos, K. Z., Wiggins, A., Gray, M. L., & Williams, A. (2017). *Crowdsourcing as a Tool for Research: Implications of Uncertainty Proceedings of the 2017 ACM Conference on Computer Supported Cooperative Work and Social Computing*, Portland, Oregon, USA. https://doi.org/10.1145/2998181.2998197

Lifshitz-Assaf, H. (2017). Dismantling knowledge boundaries at NASA: The critical role of professional identity in open innovation. *Administrative Science Quarterly*, *63*(4), 746–782. https://doi.org/10.1177/0001839217747876

Lind, F., Gruber, M., & Boomgaarden, H. G. (2017). Content analysis by the crowd: Assessing the usability of crowdsourcing for coding latent constructs. *Communication Methods and Measures*, *11*(3), 191–209. https://doi.org/10.1080/19312458.2017.1317338

Lukyanenko, R., Parsons, J., Wiersma, Y. F., & Maddah, M. (2019). Expecting the unexpected: effects of data collection design choices on the quality of crowdsourced user-generated content. *MIS Q.*, *43*(2), 623–648. https://doi.org/10.25300/misq/2019/14439

Majchrzak, A., & Malhotra, A. (2013). Towards an information systems perspective and research agenda on crowdsourcing for innovation. *The Journal of Strategic Information Systems*, *22*(4), 257–268. https://doi.org/10.1016/j.jsis.2013.07.004

Mazumdar, S., Wrigley, S., & Ciravegna, F. (2017). Citizen science and crowdsourcing for earth observations: An analysis of stakeholder opinions on the present and future. *Remote Sensing*, *9*. https://doi.org/10.3390/rs9010087

Michel, F., Gil, Y., Ratnakar, V., & Hauder, M. (2015). *A virtual crowdsourcing community for open collaboration in science processes*. *Proceedings of Twenty-First Americas Conference on Information Systems*, Puerto Ric https://www.isi.edu/~gil/papers/michel-etal-amcis15.pdf

Moher, D., Liberati, A., Tetzlaff, J., Altman, D. G., & The, P. G. (2009). Preferred reporting items for systematic reviews and meta-analyses: The PRISMA statement. *PLOS Medicine*, *6*(7), e1000097. https://doi.org/10.1371/journal.pmed.1000097

Nov, O., Arazy, O., & Anderson, D. (2010). *Crowdsourcing for science: Understanding and enhancing SciSourcing contribution*. ACM CSCW 2010 Workshop on the Changing Dynamics of Scientific Collaborations, http://faculty.poly.edu/~onov/Nov%20Arazy%20Anderson%20CSCW%202010%20workshop.pdf

Nov, O., Arazy, O., & Anderson, D. (2011). *Dusting for science: Motivation and participation of digital citizen science volunteers*. https://doi.org/10.1145/1940761.1940771

OECD. (2002). *Frascati Manual 2002*. https://doi.org/10.1787/9789264199040-en

OECD. (2015). Making Open Science a Reality, *OECD Science, Technology and Industry Policy Papers 25, OECD Publishing*.

Olson, D., & Rosacker, K. (2013). Crowdsourcing and open source software participation. *Service Business, 7*. https://doi.org/10.1007/s11628-012-0176-4

Owen, R., Macnaghten, P., & Stilgoe, J. (2012). Responsible research and innovation: From science in society to science for society, with society. *Science and Public Policy, 39*, 751–760. https://doi.org/10.1093/scipol/scs093

Pan, S. W., Stein, G., Bayus, B., Tang, W., Mathews, A., Wang, C., Wei, C., & Tucker, J. D. (2017). Systematic review of innovation design contests for health: spurring innovation and mass engagement. *BMJ Innovations, 3*, 227–237. https://doi.org/10.1136/bmjinnov-2017-000203

Parrick, R., & Chapman, B. (2020). Working the crowd for forensic research: A review of contributor motivation and recruitment strategies used in crowdsourcing and crowdfunding for scientific research. *Forensic Science International: Synergy, 2*, https://doi.org/10.1016/j.fsisyn.2020.05.002

Petersen, S. (2013). *Crowdsourcing in design research—Potentials & limitations. Proceedings of International Conference on Engineering Design*, ICED13, 19-22 AUGUST 2013, Sungkyunkwan University, Seoul, Korea, https://www.designsociety.org/publication/34839/Crowdsourcing+in+design+research+-+Potentials+%26+limitations

Poliakoff, E., & Webb, T. (2007). What factors predict scientists' intentions to participate in public engagement of science activities? *Science Communication, 29*. https://doi.org/10.1177/1075547007308009

Riesch, H., & Potter, C. (2014). Citizen science as seen by scientists: Methodological, epistemological and ethical dimensions. *Public Understanding of Science, 23*(1), 107–120. https://doi.org/10.1177/0963662513497324

Riesch, H., Potter, C. and Davies, L. (2013). Combining citizen science and public engagement: The open airlaboratories programme. *Journal of Science Communication*, 12(3), 1–19.

Rotman, D., Preece, J., Hammock, J., Procita, K., Hansen, D., Parr, C., Lewis, D., & Jacobs, D. (2012). *Dynamic changes in motivation in collaborative citizen-science projects*. https://doi.org/10.1145/2145204.2145238

Sauermann, H., & Franzoni, C. (2015). Crowd science user contribution patterns and their implications. *Proceedings of the National Academy of Sciences of the United States of America, 112*. https://doi.org/10.1073/pnas.1408907112

Sauermann, H., Franzoni, C., & Shafi, K. (2019). Crowdfunding scientific research: Descriptive insights and correlates of funding success. *PLOS ONE, 14*(1), e0208384. https://doi.org/10.1371/journal.pone.0208384

Sauermann, H., Vohland, K., Antoniou, V., Balázs, B., Göbel, C., Karatzas, K., Mooney, P., Perelló, J., Ponti, M., Samson, R., & Winter, S. (2020). Citizen science and sustainability transitions. *Research Policy, 49*(5), 103978. /https://doi.org/10.1016/j.respol.2020.103978

Savio, L., Prainsack, B., & Buyx, A. (2016). Crowdsourcing the human gut. Is crowdsourcing also 'citizen science'? *Journal of Science Communication*, 15(03), 1–16.

Scanlon, E., Woods, W., & Clow, D. (2014). Informal participation in science in the UK: Identification, location and mobility with iSpot. *Journal of Educational Technology and Society, 17*, 58–71.

Scheliga, K., Friesike, S., Puschmann, C., & Fecher, B. (2016). Setting up crowd science projects. *Public Understanding of Science, 27*(5), 515–534. https://doi.org/10.1177/0963662516678514

Schlagwein, D., & Daneshgar, F. (2014). *User requirements of a crowdsourcing platform for researchers: Findings from a series of focus groups*. PACIS.

See, L., Mooney, P., Foody, G., Bastin, L., Comber, A., Estima, J., Fritz, S., Kerle, N., Jiang, B., Laakso, M., Liu, H.-Y., Milcinski, G., Niksic, M., Painho, M., Pődör, A., Olteanu Raimond, A.-M., & Rutzinger, M. (2016). Crowdsourcing, citizen science or volunteered geographic information? The current state of crowdsourced geographic information. *ISPRS International Journal of Geo-Information, 5*. https://doi.org/10.3390/ijgi5050055

Sekaran, U., & Bougie, R. (2010). *Research methods for business A skill-building approach*. Wiley.

Shank, D. (2015). Using crowdsourcing websites for sociological research: The case of Amazon Mechanical Turk. *The American Sociologist, 47*. https://doi.org/10.1007/s12108-015-9266-9

Sheehan, K. B. (2018). Crowdsourcing research: Data collection with Amazon's Mechanical Turk. *Communication Monographs, 85*(1), 140–156. https://doi.org/10.1080/03637751.2017.1342043

Shirk, J. L., Ballard, H. L., Wilderman, C. C., Phillips, T., Wiggins, A., Jordan, R., McCallie, E., Minarchek, M., Lewenstein, B. V., Krasny, M. E., & Bonney, R. (2012). Public participation in scientific research: A framework for deliberate design. *Ecology and Society, 17*(2), 29. https://doi.org/10.5751/ES-04705-170229

Smith, K. L., Ramos, I., & Desouza, K. C. (2015). Economic resilience and crowdsourcing platforms. *JISTEM—Journal of Information Systems and Technology Management, 12*, 595–626.

Steelman, Z., Hammer, B., & Limayem, M. (2014). Data collection in the digital age: innovative alternatives to student samples. *MIS Quarterly, 38*, 355–378. https://doi.org/10.25300/MISQ/2014/38.2.02

Stewart, N., Chandler, J., & Paolacci, G. (2017). Crowdsourcing samples in cognitive science. *Trends in Cognitive Sciences, 21*. https://doi.org/10.1016/j.tics.2017.06.007

Strang, L., & Simmons, R. K. (2018). *Citizen science: Crowdsourcing for systematic reviews*. U. o. https://www.thisinstitute.cam.ac.uk/wp-content/uploads/2018/06/THIS-Institute-Citizen-Science-Crowdsourcing-for-systematic-reviews-978-1-9996539-1-0.pdf

Stritch, J. M., Pedersen, M. J., & Taggart, G. (2017). The Opportunities and Limitations of Using Mechanical Turk (MTURK) in Public Administration and Management Scholarship. *International Public Management Journal, 20*(3), 489–511. https://doi.org/10.1080/10967494.2016.1276493

Tight, M. (2018). Higher education journals: Their characteristics and contribution. *Higher Education Research & Development, 37*(3), 607–619. https://doi.org/10.1080/07294360.2017.1389858

Tranfield, D., Denyer, D., & Smart, P. (2003). Towards a methodology for developing evidence-informed management knowledge by means of systematic review. *British Journal of Management, 14*(3), 207–222. https://doi.org/10.1111/1467-8551.00375

Uhlmann, E. L., Ebersole, C. R., Chartier, C. R., Errington, T. M., Kidwell, M. C., Lai, C. K., McCarthy, R. J., Riegelman, A., Silberzahn, R., & Nosek, B. A. (2019). Scientific Utopia III: Crowdsourcing science. *Perspectives on Psychological Science, 14*(5), 711–733. https://doi.org/10.1177/1745691619850561

Wang, G., & Yu, L. (2019). The game equilibrium of scientific crowdsourcing solvers based on the hotelling model. *Journal of Open Innovation: Technology, Market, and Complexity, 5*, 89. https://doi.org/10.3390/joitmc5040089

Watson, D., & Floridi, L. (2016). Crowdsourced science: Sociotechnical Epistemology in the e-Research Paradigm. *SSRN Electronic Journal*. https://doi.org/10.2139/ssrn.2914230

Wechsler, D. (2014). Crowdsourcing as a method of transdisciplinary research–Tapping the full potential of participants. *Futures, 60*. https://doi.org/10.1016/j.futures.2014.02.005

Westra, H.-J., Jansen, R. C., Fehrmann, R. S. N., te Meerman, G. J., van Heel, D., Wijmenga, C., & Franke, L. (2011). MixupMapper: correcting sample mix-ups in genome-wide datasets increases power to detect small genetic effects. *Bioinformatics, 27*(15), 2104–2111. https://doi.org/10.1093/bioinformatics/btr323

Wexler, M. (2011). Reconfiguring the sociology of the crowd: Exploring crowdsourcing. *International Journal of Sociology and Social Policy*, *31*, 6–20. https://doi.org/10.1108/01443331111104779

Wheat, R. E., Wang, Y., Byrnes, J. E., & Ranganathan, J. (2013). Raising money for scientific research through crowdfunding. *Trends in Ecology & Evolution*, *28*(2), 71–72. https://doi.org/10.1016/j.tree.2012.11.001

Wiggins, A., & Crowston, K. (2011). *From conservation to crowdsourcing: A typology of citizen science*. https://doi.org/10.1109/HICSS.2011.207

Williams, C. (2013). Crowdsourcing research: A methodology for investigating state crime. *State Crime Journal*, *2*, 30–51. https://doi.org/10.13169/statecrime.2.1.0030

9 Tools and Technology Assisting Accreditation in Engineering Education

Prashant Gupta, Trishul Kulkarni, and Bhagwan Toksha
Maharashtra Institute of Technology, India

CONTENTS

9.1 INTRODUCTION

As the British ruled over the commonwealth countries for decades, like other things, the system for higher education in countries like India has been largely based on their legacy of higher education. The regulatory mechanisms have kept the system functioning to be satisfactory without any exceptional qualitative expansion. However, elevation in terms of higher education standards has been difficult

DOI: 10.1201/9781003102298-9

to achieve. India, in particular, has been one of the world's largest higher education systems with around 967 university-level institutions (UGC, 2020) and incorporating over 39,931 institutions (Statistica Inc., 2019). The students were reportedly around 2.94 million who are being taught by around 0.15 million faculties (UGC, 2017). The spread of relevant age group coverage is nothing in comparison to enrollments in undergraduate courses. The quality of education has to be at its best for the students under the smaller umbrella of higher education as it is crucial to the nation. The affiliating universities convene over the courses to study and hold examinations on behalf of the university. This is carried out on the basis of a common syllabus for all institutes affiliated to it and awards the degrees on the basis of the results achieved by the students. These affiliated institutes take the charge of undergraduate education and this system links the institute and affiliating universities. The academic leadership arranged to affiliated institutions have come under severe scrutiny and disparagement due to individual universities having even more than 500 affiliated institutions under them has led to ignorance of a set of conditions that are needed to ensure the required quality of education. Also, the rise of institutions, especially the private ones which are not of academic repute across the country, has increased due to the same. Although the government-run institutions are not very high in number, around 70% of the current institutions are established and managed privately by societies or trusts to better the outreach of education in our country. The privately established institutes before the 1980s are known as grant-in-aid institutes with 95% or more financial support by the state government. The privately established institutes after 1980 are known as self-financed institutes and function majorly on student fees to comply with government and affiliating university regulations. An increase in these self-financed institutions with high fee structures, substandard facilities in terms of teaching and infrastructure, etc., has added pressure on affiliating universities. This has raised the question of "value for money" in higher education, thus giving rise to a need for an effective mechanism for assurance of quality in education. The appointment of an autonomous body was considered to be an appropriate strategy for dealing in this matter where accreditation comes into the picture (Martin & Stella, 2007).

Accreditation is a continual process of quality assurance, wherein a program/course in an institution that is approved by a government regulatory body is appraised upon critical review. This is done to ascertain that the course offered by the institution, in particular, with respect to continuity in meeting and/or exceeding the standards prescribed by the regulatory authority from time to time fulfills a set of proposed standards for the program. As a consequence, in India, the University Grants Commission (UGC) established the National Assessment and Accreditation Council (NAAC) and National Board of Accreditation (NBA) in 1994 within their role for maintenance and promotion of education standards by UGC and All India Council for Technical Education (AICTE). The methodology is in alignment with the international trend of self-evaluations to go before the peer review process for assessment of various criteria. While NAAC is for institutional accreditation, NBA is for the program offered by the institution.

9.2 MECHANISM OF ACCREDITATION

The quality assurance in engineering accreditation comes from the professional practices defined for educational functioning. The pedagogy designed needs the inputs from students, academic quality system, the academic operational environment, academic program, and most importantly from the industries/experts. The expert opinion from the best in the trade (industry-side) forms the key basis of the accreditation philosophy. This philosophy is altogether an intangible complex to put together and comprise of abstract measurables such as Program Outcomes (POs), Program Educational Outcomes (PEOs), Program-specific Outcomes (PSOs), and so on, the terminology of which is dependent upon the accreditation agencies (AAs) across the globe. The discipline-specific and institutional learning outcomes along with societal expectations contribute toward defining graduate capabilities that are expected to be imbibed during an engineering program. There are well-laid measurables which are determined and presented to these AAs during the accreditation process in one of the two modes, i.e., online and/or onsite visit to the institute. The mechanism of accreditation with philosophical steps involved is shown in Figure 9.1.

Some of the prerequisites that are necessary before applying for accreditation bodies, such as NBA, is a minimum percentage of admissions to the applying program. While it is very difficult to judge the quality of the program just on the basis of admissions to it, it is important to have a class with enough students to have a good quality of higher education. The faculty feels motivated if he/she is teaching to a class full with students and can effectively disseminate the required course content in a confident and inspired manner (Díşlen et al., 2013). It is quite common to learn the application of knowledge, problem-solving skills, teamwork, design and development of solutions, solve complex problems, teamwork, communication, etc., while studying in the institute through engaging oneself in learning, competing among the best in the business, doing projects, forming a hypothesis, etc. However, some other aspects have been identified as an outcome to ensure the quality of higher education. For instance, lifelong learning has been termed as the life-wide, voluntary, and self-motivated pursuit of knowledge for professional as well as personal reasons. It has the capability to enhance social inclusivity, development at an individual level, active citizenship along with boosting employability and competitiveness (Ates & Alsal, 2012). Ethics and moral values are other aspects that have been explored in connection with the quality in the higher education landscape (Prisacariu & Shah, 2016). The global expansion of higher education has resulted in a shift from elite to mass. With this shift, there have been significant risks which as well studied and documented (Lynch et al., 2020). However, ethics and moral values as virtues of higher education have been less researched. There are pieces of evidence for instance, over unethical practices of individuals (Osipian, 2014), among others, which are obviously questioning the integrity of those who are involved in the teaching–learning process. This has been identified as an important aspect and hence included within the list of outcomes on which the evaluation is conducted for accreditation purposes.

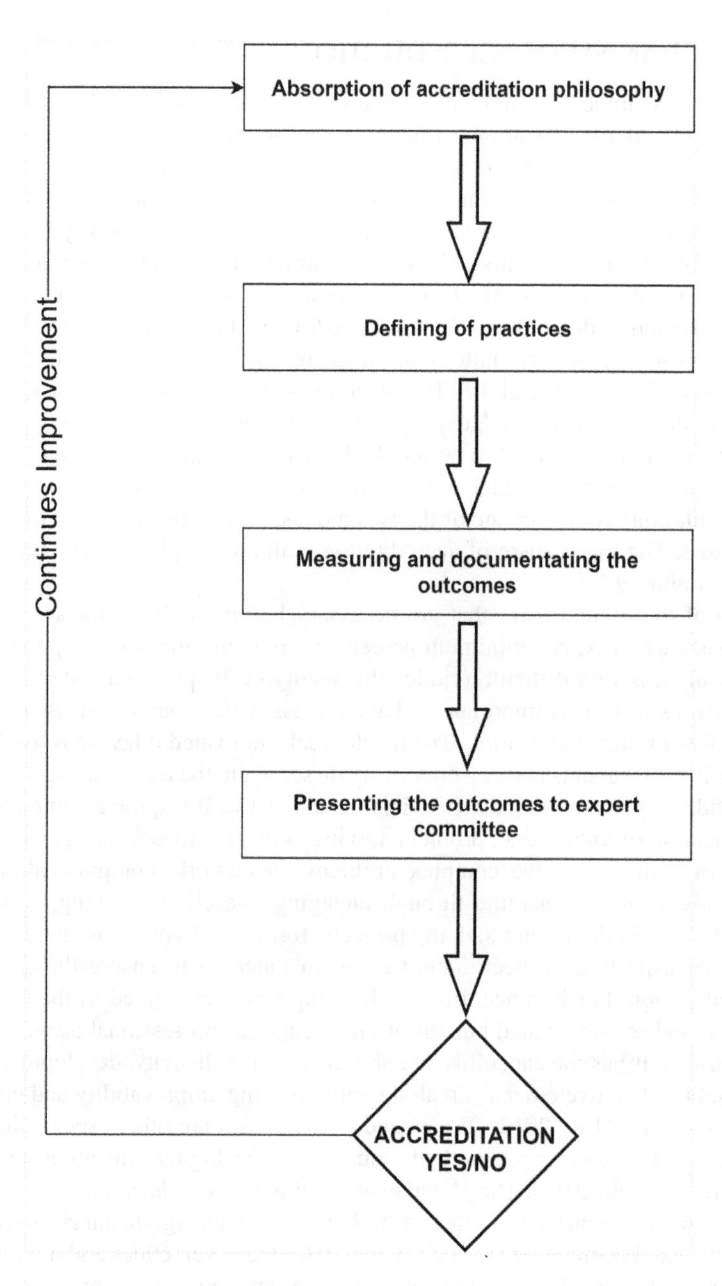

FIGURE 9.1 Mechanism of accreditation.

9.3 ACCORDS AND AGREEMENTS PERTAINING TO EDUCATIONAL ACCREDITATION

The 21st-century world is growing as a global market, overcoming all the physical barriers which have created a dependency for a runaway supply of resources in terms of raw materials, ready to use products, and human expertise. This change is not a superficial one and demands much deeper arrangements to be done. For example, the governments of China, Germany, and Japan, which had traditionally used native languages encouraged the education in the English language. At the same time, the rest of the world embraced Chinese, German, and Japanese languages for having a better connect with them. Another example is the acceptance of common SI units and measurements keeping the "foot pound second" system at a side (Jinghui Si, 2019; Kubota, 2017; Nikolantonakis, 2006). Thus, the requirements regarding modern-day competency related to engineering practices are increasing at a very fast rate. This, in turn, is demanding the global education agreement with respect to standards, best practices, accreditation processes, and mutual recognition of accredited engineering programs and agreements for defining and recognizing professional competence. The engineering professional life cycle begins with meeting a minimum standard for engineering education and further comprises steps such as observing the code of conduct and meeting the standards for professional competency. The role of these agreements is to protect the interests and enforce the rights of all the parties involved. It helps in bringing clarity to understand the expectations of them and, more importantly, agreeing upon some specific facts. From the perspective of the global workplace, and even the skill set and mobility of professional engineers is unavoidable. The world is inevitably heading toward globalization which implies that the world has become a single global market. The immediate implication of this situation is the requirement of an unhindered flow of products and technical expertise across national borders. Besides mergers and acquisitions, the other thing that can be done is the standardization of processes and practices. Keeping this scenario in mind, the mission of various accords was designed to improve technical education globally and foster the mobility of students, graduates, and professional/skilled persons (Balikaeva et al., 2018; Mikhnenko & Absaliamova, 2018).

The members from 29 countries constituted a global non-profit organization under the entitlement "International Engineering Alliance (IEA)" (*https://www.Ieagreements.org/*, 2021). This alliance functions for recognizing engineering educational qualifications and professional competence with guidelines officiated in the form of seven international agreements. The aim of these agreements is to establish and enforce the benchmarks and standards pertaining to the engineering education and expected competence for engineering practice accepted globally. The Washington Accord (WA) was the first constituent of the IEA accepted and enforced in 1989. The main focus of this agreement was the mutual recognition among its signatories of accredited educational programs. The WA was designed in order to provide the educational foundations for professional engineers. The next two accords came into existence in the form of the Sydney Accord and Dublin Accord, with the objective of programs providing the educational foundation for engineering technologists and concerning with engineering technicians, respectively. Figure 9.2 depicts the structure of IEA along with the accords and agreements with the timeline.

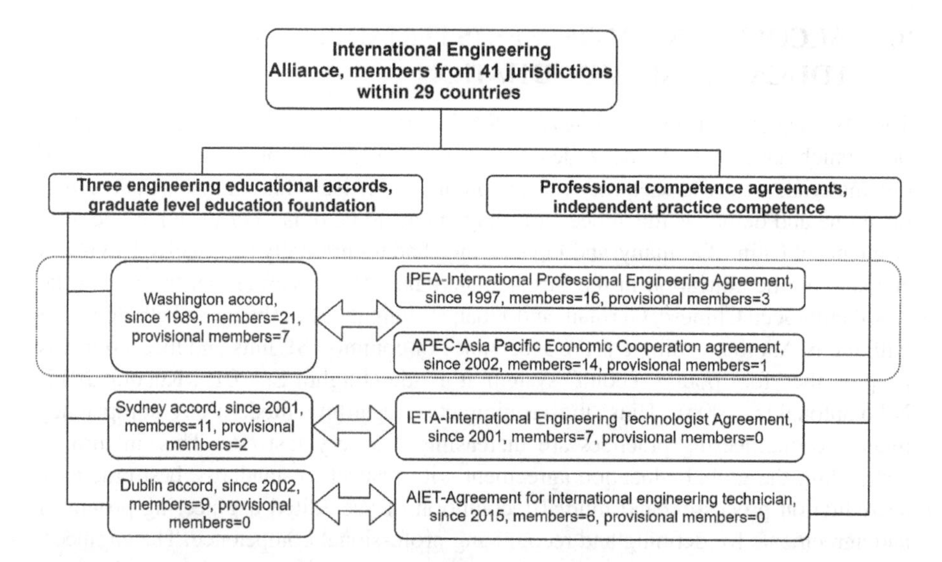

FIGURE 9.2 The structure of International Engineering Alliance (IEA) along with the three accords and agreements with the timeline.

9.3.1 The Washington Accord

The WA is an initiative to standardize engineering competencies in terms of learning outcomes and skills attained by the engineering graduate in the OBE environment. The initial thought behind forming the signatory group was to include the countries having similar existing engineering academic programs and accreditation processes which will eliminate the need for any significant reforms. One of the major factors contributing toward it is definitely the medium of instruction. Thus, the WA was firstly initiated by six countries predominantly having English as their native language. It comprised of countries namely: Australia, Canada, Ireland, New Zealand, the United Kingdom, and the United States of America (USA). The WA covers undergraduate accredited engineering programs within the WA countries for mutual recognition by all members of the WA. The WA is in full operation and is a multilateral agreement between AAs in the signatory countries that have chosen to collaborate with an aim to assist the mobility of engineering professionals. These AAs functioning in their respective countries are responsible for accreditation/recognition of regional engineering qualifications within their jurisdictions. The ease of mobility of engineering professionals sometimes may be called a brain drain in a negative sense. This mobility of engineering professionals from one country to another has been a topic of discussion for many decades. The argument against the mobility of engineers goes in the direction that the foreign engineers are using up the vital and scarce human talent to provide the countries they visit with highly skilled human resources at a very low cost. The use of a highly specialized resource is essential to the economic and social development process of the native country. The argument that goes in favor of the mobility of engineering professionals is that the movement of highly educated individuals is important since it results in a more efficient allocation of all important human resources. This argument considers that having the highly trained engineers fixed in boundaries would result in overall lower productivity. The exchange and not the immigration of highly skilled manpower from one country to another will significantly contribute to the development potential.

The obligations and benefits of WA are as follows:

1. The participant in the accord accepts the fact that the criteria, policies and procedures used by fellow participants in the accord are comparable.
2. The accreditation decisions rendered by the participants in the accord are acceptable to the other signatories.
3. Each participant will practice ensuring that the AAs responsible for licensing professional engineers to practice in its region accept the substantial equivalence of engineering academic programs accredited by the signatories to this agreement.
4. Participants agree to exchange information on their respective criteria, policies, and procedures in order to encourage the implementation of best practices.
5. The observation made by other AAs is encouraged outside the formal monitoring visits.
6. A single set of criteria in detail may/may not meet; it will be accepted as an education base for overall POs. This relates to the output of degree programs rather than the detailed internal structures of the program.
7. The identity of a graduate will be a graduate of an accredited program.
8. Equal rights and privileges to graduates from signatory countries.
9. The graduates have met the professional and academic essentials in their countries if the country is under the accord and the institute has got the accreditation.
10. The ease in the mobility of professional engineers.

9.3.2 The Bologna Process

The Bologna Process (BP), with common expectations, was designed to lead toward the creation of a European Higher Education Area (EHEA) for higher education policy reforms and overcome obstacles in higher education (*https://ehea.Info/*, 2021). The establishment of BP took place in 1999 with a consortium of 29 European countries. As of date, the consortium has expanded with 49 full members covering the entire European continent till date. The BP evolved as a channel of dialogue and platform for taking collective diplomatic decisions across Europe regarding higher education reforms. It also covers all academic programs including engineering at the undergraduate and postgraduate degree levels.

The objective of BP was to let higher education systems work coherently across Europe. The targets of the BP in the context of Europe were as follows:

1. Facilitation of student and staff mobility by mutual recognition of study periods and qualifications earned
2. Making higher education more inclusive and accessible

3. Making higher education more attractive and competitive worldwide by enhancing the quality and relevance of teaching/learning
4. Cross-border academic cooperation

All participating countries under BP agreed upon the introduction of a three-cycle higher education system consisting of bachelors, masters, and doctoral studies; ensuring the mutual recognition of qualifications and learning periods abroad completed at other universities; and implementing a system of quality assurance, to strengthen the quality and relevance of learning and teaching.

While WA and BP are the international agreements about accreditations providing guidelines about engineering education standards, one of the agencies, ABET, i.e., Accreditation Board for Engineering and Technology has achieved a greater perspective as an international AA. The agencies like NBA in India and Engineers Australia in Australia function within its national territories, whereas the ABET is a global AA which was formulated in America and expanded beyond America worldwide. The ABET assumes the role of being a non-governmental agency that accredits programs in Science, Technology, Engineering, and Mathematics (STEM) with an approach of continuous improvement. The ABET accreditation is done for the programs rather than the institutes which claims the assurance that an institutes program has met the quality standards of the profession. The accreditation is done for post-secondary programs barring PhD degrees within the institutions already recognized by regional or national education authorities. As of date, the ABET accreditation is awarded after the online assessments to more than 4,000 programs at spreading over 846 institutes belonging to 41 countries.

The comparative points in WA and BP, as pointed out by Kasuba et al. (Kasuba & Ziliukas, 2004), are that the WA involves the undergraduate portion of the regular undergraduate–postgraduate–PhD sequence and the BP addresses the short and long undergraduate and postgraduate programs, degree designations, substantial variations in the required number of credits and definition of credits for earning a degree. There could be variations in WA and BP regarding flexibility and scope, however, they still have an overall complementary nature toward each other. The principle aims of both these accords lead to higher quality levels in accreditation and engineering education remains the same. While the WA and BP do not cover the national or international licensing of engineers, the national licensing policies are designed by the participating countries and rely on their respective accreditations. The need to get recognized on the international platform, making the graduates moveable/acceptable to other countries, and desire to highlight the quality of education imparted to the graduates are the driving force behind becoming a part of an accord at the national level and getting accredited at the institutional level. The role of learning outcomes, assessment, and continuous improvement was reported to be more critical as compared to curricular specifications. The international accords among engineering accrediting bodies were also found helpful in facilitating the mobility and globalizing the engineering profession (Prados et al., 2005).

9.4 INDIAN CONTEXT OF ASSESSMENT AND ACCREDITATION

There exists a disparity of high-quality global engineering students produced in India and few developed countries, for example, engineering students from the USA. The doubts regarding the quality of engineering graduates continue to challenge the engineering graduates from China and India. A first-hand study on the implementation of OBE in engineering institutes across India was carried out by Komives et al. (Komives, 2015). The challenges in terms of philosophical differences between Indian and USA accreditation that needs to be addressed are curriculum, faculty, and facilities. This is one of the prominent reasons behind the prospect of substantial unemployment, despite high corporate demand for their services. However, the engineers from developed countries also confront problems in its continued ability to attract and retain top engineering talent from abroad due to various problems related to immigration uncertainties and growing economic opportunities in their native country. The quality of graduates which is going to play a decisive role in innovation and entrepreneurship needs to be addressed at all policy levels. The accreditation activity increases the active engagement in their learning; ensures higher and qualitative interaction with instructors with immediate and meaningful feedback on their work from the instructor. There are certain advantages of participation in these international accords, e.g., it opens the opportunities regarding studying abroad, travel opportunities through international students exchange programs, more involvement in engineering design competitions, emphasis on programs through openness to diverse ideas and people.

Accreditation is a statement of meeting the minimum requirements which are voluntary and the programs which are known for academic excellence may not choose to be accredited. The process of accreditation is understood as a formal recognition of an educational program by a designated agency based on an assessment of quality for a given duration. In this process, representative committee from the designated agency is invited which examines an educational program to ensure that it is meeting minimum standards established by experts in the field and industry through an evaluation process. This process results in terms of two outputs either the said program is accredited or not accredited. In the Indian context, the rules of NBA were constituted under "Section 10 (u) read with section 10 (r) of the All-India Council for Technical Education (AICTE) act, 1987". The regulations of 2015 onward require that all engineering programs attain the POs and demonstrate that they are following continuous improvement policies. The undergraduate and post-graduate programs offered at Higher Education Institutions (HEIs) need to get NBA accredited by 2022. The representation of India in WA is NBA. India holds the status of being permanent signatory status in WA since June 2014. The *Tier-1* institutions in India which are eligible to get recognized under WA are the Indian Institutes of Technology (IITs), Indian Institute of Science (IISc), Indian Institute of Information Technology Design and Manufacturing (IIITDM), Indian Institutes of Science Education and Research (IISERs), Indian Institutes of Information Technology (IIITs), National Institutes of Technology (NITs), and the central, state, private, Deemed-to-be universities and autonomous institutions.

National Policy in Education (1986) recommended the existence of an agency that will carry out the assessment and accreditation of higher education institutes, in response to which UGC established an autonomous institute as NAAC in 1994 (*http://www.naac.gov.in/*, 2021). The aim and objectives of NAAC were set to not only assess and accredit higher education institutes but also helping them to work continuously to improve the quality of education. NAAC is a member of the International Network for Quality Assurance Agencies in Higher Education (INQAAHE). INQAAHE is an international quality assurance body established in 1991 that functions in 140 countries with a motive to control educational quality in around 140 countries. A major shift in NAAC focus came in 2017 when the NAAC framework was awarded the features of being ICT enabled, objective, transparent, scalable, and robust. All general higher education programs were required to attain the POs and PSOs. The student-centric approach along with outcome-based education (OBE) was embraced and it was made mandatory to plan and conduct all teaching/learning processes to facilitate the students to attain well-defined and measurable outcomes i.e., OBE. The quality of higher education is attributed to inspections and audits by state governments, the regulating support toward affiliated institutes from universities, performance appraisal by the apex bodies such as AICTE and UGC, and other funding agencies (Stella, 2002). The shift of emphasis of awarding degree is valid academic/employment purposes for the assurance of quality in the context of India was reported by Sinha et al. (Sinha & Subramanian, 2013). The article concludes the educational institutes could gain excellence in academics through the evaluations carried out internally and externally, endorsements, and sustenance initiatives. The expansion of ABET, USA, accreditation, and the WA worldwide were projected as an opportunity for HEI's in India while the Indian context of accreditation and recognition mediated through industrial quality assurance was related to survival and sustainability of higher educational institutes by Natarajan et al. (Natarajan, 2000). The parameters for accreditation of undergraduate degree programs in view of India becoming a member of WA along with the critical points to be considered are exhaustively analyzed in a study by (Banthiya N K, 2012). The study brought out the important features and made suggestions for successful implementation. The student and faculty contribution were claimed to be the backbone of accreditation. The era prior to India becoming the permanent signatory of WA and the role of industry and corporate sector to counter the un-employability of Indian engineering graduates was elaborated by Sthapak et al. (Sthapak, 2012).

9.5 FACTORS DETERMINING QUALITY IN HIGHER EDUCATION

As said earlier, the economic success of a nation is interlinked with the quality of education systems in place and the most influential aspect of manufacturing is human capital in the form of skills, knowledge, creativity, and morals of the individuals in the society. Since the early 2000s, higher education establishments have been constantly put under the pump due to various pressures across the globe such as growing significance of knowledge-led economies with higher education at the core of nation competitiveness plans. In the modern business world, the higher education establishments are being looked at as economic engines by the government policymakers to ensure knowledge creation via the use of innovation along with continual tutoring of the people involved. The higher education quality policies with respect to tertiary education is increasingly gaining importance on national agendas as it is a major driving force of economic competitiveness in today's knowledge-driven global economy. From a societal perspective, there is a growing need to raise high-level employment skills for sustaining a globally competitive research base and to expand the spread of knowledge (Pavel, 2012).

The quality of education concept has been termed vague and controversial as a large number of stakeholders with divergent interests are involved with their own versions of quality (Cheong Cheng & Ming Tam, 1997). The controversy is due to students being considered and treated as customers (Cuthbert, 2010; Guilbault, 2016; Obermiller et al., 2005). The concept of quality in education in the reported literature is very much like the quality of service wherein a set of parameters which are input, output, or process-related are recognized to outline the same (Quinn et al., 2009; Teeroovengadum et al., 2016). The factors affecting the quality of education are summarized as under:

9.5.1 Academic Standard

The academic standard of an institute largely hangs on various parameters such as faculty profile, curriculum design flexibility, curriculum dynamism as per current societal requirements, etc. The focus has been on these factors in spite of the unprecedented growth of higher education (Bovill & Woolmer, 2019). The shortfall in terms of faculty and irrelevant curriculum is considered to be at the top of the ladder in terms of factors responsible for the uncertain quality of higher education in the Indian context (Jagadeesh, 2000).

9.5.2 Industry Connections

Of all the purposes served through higher education, one of the very important ones is to develop problem-solving approach and skill development (Changwong et al., 2018; Romero et al., 2017). All in all, the approach is toward developing managers for tomorrow. However, the literature suggests that institutions are not able to meet the academic and social needs along with demands regarding the same (Alexander et al., 2019). For this to happen, connect with the industry plays a very important role. The institutions should be able to understand the deliverables/objectives with respect to modern-day requirements and that is where this connection plays a very important role in the exchange of the required information (Anand et al., 2021). If the industry linkage is good, it ensures continues inputs to enable effective changes in the curriculum design and student training. Also, student internships, in-plant industry visits, and joint projects can go a long way in practical exposure and skill development for the students (Ghobakhloo & Fathi, 2019; Shatunova et al., 2019). The arrangements of guest lectures from industry experts can help them in the augmentation of the latest updates and technical know-how (Pratibha et al., 2021). It can potentially affect other factors such as research and development, students' absorption in the industry, and boosting the financial wellbeing as the institution on a whole.

9.5.3 Organization Structure and Practices

Organizational coordination has been reported as important for effective quality in higher education (Oza & Parab, 2011). Also, practices such as providing freedom given to the faculty in terms of teaching and research, motivation, non-teaching staff efficiency, managing talent and knowledge, organizational rules and regulation, work environment and culture, organizational vision and mission, leadership, employee focus, financial perspective, strategic planning and internal processes (Mohammed et al., 2016).

9.5.4 Research and Consultancy

The role of research/consultancy in understanding, testing, or developing new concepts is crucial in the intellectual growth of students/faculty (Cullen et al., 2003). The process helps in accumulating substantive knowledge and findings to share them through the right channels for the generation of subsequent questions for further inquiries. It can be done through a case study, research papers, thesis, and report writing along with consultation and training for the industries. However, research and consultancy are not everyone's job as a lot of institutions with higher education fail to impress in these aspects due to a variety of factors including poor infrastructure, substandard faculty and students, insufficient financial support, publication pressure, etc.

9.5.5 Accreditation

As mentioned earlier, it encourages the promotion of best practices in higher education via a benchmarking process (Sahay & Thakur, 2007). There have been factions between HEIs on the basis of quality and accreditation from a third party national or international recognized agency is the way forward. While the Indian bodies consist of NBA and NAAC, international bodies such as INQAAHE with European Association for Quality Assurance in Higher Education (ENQA), National Agency for Higher Education (NAHE), Finnish Higher Education Evaluation Council (FINHEEC), Länder, Hungarian Accreditation Committee (HAC), Netherlands Accreditation Organization (NAO), Danish Centre for Quality Assurance and Evaluation of Higher Education and most importantly Accreditation Body for Engineering and Technology (ABET) are responsible for either accreditation or contribute toward the improvement of higher education quality. The effects of accreditation are on all other factors and hence it is interlinked with them.

9.5.6 Placements

One of the most important parameters that drive student decisions regarding selecting higher educational institutions is placements and the average salary offered. This is due to the students joining these institutes in search of better jobs than offers at undergraduate level institutions. However, the search for a better job is rather elusive and need an elaborate discussion. Kettis et al. discussed how placements have been an underused vehicle for enhancing quality in higher education. They articulated how placements can be made more worthy for the students along with improvement in the quality of teaching–learning at HEIs. They suggested that it also led to creating a work culture more conducive to learning. They also highlighted placements contribution toward organizational development and concluded that interactions of HEIs with employers benefitted both stakeholders, especially for academics by developing insights into practice which may improve the current level of teaching that can be aided by taking help from real-life examples that may trigger student motivation. It certainly aids the curriculum design as well (Kettis et al., 2013). As there has been an enormous increase in the importance of placements, sanity has to prevail in terms of not converting the institutions into some sort of placements agencies instead of knowledge disseminators (Hay, 2008).

9.5.7 Abiding by Regulatory Bodies

The rules and regulations laid down by the regulatory bodies, such as UGC, AICTE, Indian Council for Agricultural Research (IGCAR), etc., are immensely important for the success and continual improvement of quality in a higher educational institution (Jha & Kumar, 2012; Kumar & Dash, 2011). As they have been put to place with immense thought to ensure the minimum quality of higher education, abiding by them will prove to be beneficial for all the stakeholders including the students. However, there are several institutions that do not abide by the set of regulations even after their approval due to various reasons such as expenditure involved in having the required resources (Jagadeesh, 2000).

9.5.8 Financial Resources

For any Indian institution to be world known in terms of quality on offer, availability of funds is the biggest hurdle (Rao & Hans, 2011). Depending upon the category of the institution, the funds may be from state/central government, student fees, project grants provided by government bodies, industry projects, and consultancy support, etc. The dynamics of such resources are totally based on how the institution is run. The government institutions have to pass through many layers of bureaucracy to obtain funds for development, expansion, etc. On the other hand, the privately-owned institutions are totally oriented toward the profit-making cycle which is majorly based on admissions. More often than not, the investments related to development and expansion are a result of recent success the institution might have had in terms of profits by the institute.

9.5.9 Leadership

The role of a leader toward maintaining and improving quality is an important one as the formal head of the institution is responsible for establishing the core values with respect to quality. Also, a leader is the key decision-maker for ensuring institutional commitment and investment toward the common goal of ensuring and improving quality in higher education (Shiramizu & Singh, 2007). The role of leadership in creating a vision, making and communicating policies, and deploying strategy throughout higher education establishment has been reported (Davies et al., 2001). It has also led to an improvement in higher education faculty. Skender reported a managerial framework for leadership management of newly established universities by promoting a total quality educational concept as the changes in well-established institutions may be very tough to bring about (Bruçaj, 2019). A lean six sigma leadership model has been tried by Lu et al. and a conceptual framework based upon leadership, statistical thinking, continues change, and improvement (Lu et al., 2017).

9.5.10 Co-Curricular Activities

The recruiters today are keener to hire people with all-round skills instead of just textbook knowledge and problem-solving skills. The rise of co-curricular activities in developing discipline, commitment, and tenacity is well known. These values are seldom taught in the class or received from the books. The critical thinking, reasoning, ethics, quantitative analysis skills, etc. that are more training-oriented, the values learned from co-curricular activities are key skills for success in terms of mitigating unknown challenges and complexities in life. These activities bring together students from different races, religions, ethnicity, and possibly cultures on the same platform for industry-academia interactions and help in grooming their personality and holistic development as human beings.

9.6 TOOLS UTILIZED FOR ACCREDITATION AND EVALUATION

The assessment quality in an educational system drives the teaching–learning process. A poorly designed assessment process can hamper the quality of the teaching–learning process rather than improving it. The assessment of learning outcomes should not merely be an activity for accreditation compliance; rather, findings in the student learning outcome assessment should be proactively used as feedback for improving the teaching–learning process. Though it is a well-accepted fact, assessment, and documentation of student learning outcomes and using it as evidence for improving student and institutional performance is a complicated and time-consuming process. If done properly, the student learning outcome assessment has the potential to be a very effective tool for improving student performance. To fulfill the requirements of accreditation, many institutions adhere to the culture of compliance by separating assessment of outcomes from core teaching–learning process or in some cases by even outsourcing it (Matters, 2016).

The accreditation process based on OBE necessitates various reports related to criteria as listed in Table 9.1. The AAs in various countries have divided the accreditation process into a number of criteria that an educational program must satisfy in order to get accredited. For example, Accreditation Board for Engineering and Technology (ABET) has specified seven general and one program-specific criterion whereas the NBA has specified ten criteria. These criteria can be classified into four categories as shown below (*Accreditation Criteria & Supporting Documents | ABET*, n.d.).

TABLE 9.1
Typical accreditation requirements

Sr. No.	Category	Details	ABET criteria No.	NBA Criteria No
1	Student assessment and evaluation	Evaluation of student progress and performance, student outcome attainment, and continuous improvement	1,3,4	3,4,7
2	Program and curriculum	Program educational objectives (PEO), program curriculum, and teaching–learning processes	2,5	1,2,8
3	Faculty information	Information about the faculty members related to their qualification, contribution, competence, service activities, professional development, and interactions with industrial and professional practitioners	6	5
4	Governance, facilities, and support	Information about the governance, infrastructure, and facilities required to provide an atmosphere conducive for learning including institutional services, financial support, and non-teaching staff	7,8	6,9,10

A program to be accredited must satisfy all the criteria mentioned in Table 9.1 with documented evidence. The entire self-assessment report (SAR) prepared by the program is lengthy and contains numerous sections. The most demanding, important, and at the same time most overwhelming criteria are related to student outcome (SO) assessment and evaluation. The Criteria 3 of ABET and Criteria 3–4 of NBA talks about the SO and performance assessment. These criteria define a set of POs, which identifies the set of abilities and skills related to various learning domains like cognitive, affective, and psychomotor. ABET defines 11 POs whereas NBA has set 12 POs to which students should attain at the end of an engineering program. The educational institutes should define the course outcomes (COs) for each course and map these COs to POs for calculations of PO attainment. The AAs require that an educational program should devise a system that will carry out and document the assessment of SOs and performance toward the attainment of POs. The assessment part is further divided into two categories like direct assessment which is based on assessment instances like student grades in assignments, quizzes, exams, and other instruments of direct assessment. The indirect assessment of POs includes course-end, program-end, alumni, employer, and parent surveys. The direct assessment data is collected by faculty members for their respective courses while indirect assessment data is gathered through various surveys from enrolled students, graduated students, alumni, parents, and employers. These surveys utilize objective questions with Likert scale assessment to collect feedback from these stakeholders regarding the program and its related elements. The final assessment of POs will be a combination of direct and indirect assessment and a standardized outcome assessment report is prepared based on analysis of PO assessment along with other parameters as per the requirements of the AA (E. Essa et al., 2010; Sabir et al., 2018).

A typical engineering curriculum offers 10 to 12 courses (including laboratory courses) per semester and a student must complete 6 to 8 semesters in an engineering program. If we consider the average intake capacity of each engineering program as 60, it leads to the generation of a tremendous amount of data for collection and processing. Traditionally, most of the institutes which are in the process of preparing themselves for the accreditation process are following a manual approach to collect and process this data (Dew et al., 2011; Shankar et al., 2013). The preparation of reports pertaining to SO assessments is a nontrivial task and thus requires burdensome documentation. The fact that there is a lack of standard procedures available for the preparation of these reports has enforced each such institute to study and generate their own formats and procedures depending on their understanding of OBE assessment (E. Essa et al., 2010; Hussain et al., 2020).

The use of modern Information and communication tools for data collection, management and analysis can make the SO assessment process much more efficient and, in turn, encourage wider faculty participation in the accreditation process. A considerable amount of time can be saved by the use of a variety of ICT tools in an accreditation process which can potentially serve as a boon for academicians which are burdened by administrative activities (Balasangameshwara, 2015; Sabir et al., 2018). The time thus saved can be implemented into various academic and research-oriented activities which can again be fruitful for scoring a number of valuable points in the accreditation process.

A variety of software tools are available to support outcomes-based continuous program improvement processes. These tools are classified into Content Management System (CMS), Assessment platforms (AP), Analytics System (AS), and Curriculum Mapping Tool (CMT). A standard learning management system (LMS) includes modules for content creation, content delivery, course administration, learning assessment, learning analytics, and curriculum mapping tools (Kaupp & Frank, 2014, p. 2). Several LMS solutions are available like Moodle, ATutor, Open edX, Chamilo, Canvas, Forma, Dokeos, Ilias, and other few in the open source category (Cavus & Zabadi, 2014) and Litmos, LearnUpon, TalentLMS, Blackboard, Google Classroom are available commercially. These LMS are fulfilling most of the requirements of educational institutes but fail to transform the data to information and reports as per the requirement of AAs (Elhassan & El-Hassan, 2019).

There are various challenges while selecting appropriate tool which will help in the accreditation process. The institute can either develop in-house applications using various technologies or can prefer a commercial solution available in the market. However, selecting a tool is still not a straightforward job as each option poses a lot of challenges. The first challenge is the unavailability of literature that will guide to choose the appropriate option. Secondly, various AAs have their own set of requirements so finding a customized tool that is best suited for one's requirements and practices is a difficult task. Building a custom tool from scratch requires in-house software engineering skills, knowledge of proper development tools, and requirements of personnel, resources, and time. On the other hand, it caters to the program's needs and is very easy to integrate into the assessment processes that are already in place. The attempts made to build a custom solution for catering to the accreditation requirements are listed in Table 9.2. One can save the time and effort required for development time by choosing a commercial application with a cost. However, the institutes are reluctant to deploy the commercial solutions as they may not address specific requirements and do not wish to involve a third party for the analysis of sensitive assessment data. Table 9.3 gives an exhaustive list of the commercial solutions available in the market.

TABLE 9.2
Various custom/in-house tools developed to assist the accreditation process

Sr. No.	Tool	Use / Scope	Reference
1	ABET Accreditation Manager	A web-based application developed to automate the documentation, analysis, and report generation for ABET criteria 3	(Shankar et al., 2013)
2	ACAT (ABET Course Assessment Tool)	A web-based application developed for streamlining and automating the existing manual course assessment process for ABET criteria 3	(Eugene Essa, 2010)
3	Accreditation Tool (ACT)	A database application developed using Microsoft access for automating the activities related to ABET criteria 3	(Sabir et al., 2018)
4	iOBE: The Integrated OBE Software	A free software providing applications to address the requirements of implementing OBE	(Mallikarjuna et al., 2020)
5	AbOut	A fully functional software designed for assessment and evaluation process for ABET accreditation	(Schahczenski & Van Dyne, 2019)
6	Engineering Education Accreditation Aided Management System (EEAAMS)	A web-based application for the standardization and systematization of Chinese engineering education accreditation.	(Sun et al., 2017)
7	OBACIS: Outcome-Based Analytics and Continuous Improvement System	Designed and implemented as a suite of applications for collection, organization, analysis, and presentation of the data required for accreditation requirements	(Ismail, 2016) (Ismail, 2017)
8	ADAMS: Accreditation Data Analysis and Management System	An online system developed to operate alongside the LMS such as Blackboard Learn and Ellucian's Banner.	(Elhassan & El-Hassan, 2019)
9	Course Assessment Tool	A web-based application to streamline the assessment data collection, performance evaluation and tracking of remedial recommendations.	(Ibrahim et al. 2015)
10	Accreditation Management System (AMS)	A web-based application to simplify tasks related to collection, analysis and report preparation, i.e., Canadian Engineering Accreditation Board requirements.	(Dew et al., 2011)

TABLE 9.3
Various commercial tools available to manage the accreditation process

Sr. No	Tool	Use / Scope	Product Website
1	Creatrix Campus	A centralized accreditation management software for student outcome assessment, feedback, and international accreditation process	https://www.creatrixcampus.com/
2	EduSys College Management Software	It provides an all-in-one OBE System as part of educational ERP system for NAAC or NBA Accreditation	https://www.edusys.co/en-in/obe-software.html
3	EvalTools (Namoun et al., 2018)	It provides a complete solution with three modules, i.e., learning management system, outcomes assessment system and academic administration system for ABET accreditation	http://www.makteam.com/index.php/evaltools-highered-iis
4	Inpods (Balasangameshwara, 2015)	It is a cloud-based platform, to capture and analyze the curriculum and student outcomes data for NBA and ABET accreditation	https://www.inpods.com/
5	IonCUDOS (Mahadevaiah et al., 2018; Sawant, 2017)	It is a cloud-based software for automation of OBE processes for NBA accreditation	http://www.ioncudos.com/
6	Ki-OBE	It is a cloud-based accreditation analytics software customized for NBA accreditation	https://www.kramah.com/nba-obe-software/
7	Linways	The complete NBA Accreditation Software	https://linways.com/
8	MasterSoft	It is an educational ERP system compliant with NBA accreditation	https://www.iitms.co.in/
9	Smart Accredit (Namoun et al., 2018)	It incorporates a learning outcomes assessment module specifically for criteria 3 of ABET accreditation	https://www.smart-accredit.com/
10	WEAVEonline (Namoun et al., 2018)	It provides various modules like assessment, accreditation, and faculty credential management with focus on the continuous improvement of all academic programs	https://weaveeducation.com/

9.7 CASE STUDIES

Ravi Shankar (Shankar et al., 2013) and his team developed a software application "ABET Accreditation Manager" to streamline the ABET accreditation process for a computer engineering program. This tool was developed to automate the documentation, analysis, and report generation for criteria 3 in ABET which requires calculating the student learning outcomes that support the PEOs. A web-based application was developed which accepts course-wise input in the form of (comma separated variables) CSV file, this data is then populated in MySQL database. The faculty members are required to add course-specific data in this application every semester.

At the end of the academic year, the graduation coordinator, which is a second-level user, identifies the graduating students in a given semester. The application will automatically publish the SO reports as required in criteria 3 with action-required items. Though authors claim that this software will help ABET faculty coordinators in engineering academics to streamline and automate the activities for criteria 3, they have not provided any evidence related to the efficiency and effectiveness of this tool.

An in-house database application named the Accreditation Tool (ACT) was developed and used at the department of electrical engineering, University of Hafr Al-Batin, Saudi Arabia for managing and analyzing assessment data for ABET accreditation purpose. Microsoft Access was used to design this application that combines relational database management with a graphical user interface (GUI) and software-development tools. Microsoft Access is available as a part of the Microsoft Office Professional Suite which is mostly available with institutes as licensed software. The reported that it is possible to develop a fully functional multi-user application for collection, maintaining, and analyzing of ABET accreditation data by using Microsoft Access. They also provided a basic version of this application with a user guide for readers. This application is designed specifically to address the ABET criteria related to SO assessments. ACT collects direct and indirect assessment data from users in the form of student grades and feedback from various stakeholders. This data is then characterized as per the ABET guidelines and in terms of the accomplishment of SOs and COs. The departmental committee can analyze this data using ACAT to identify correlating patterns and prominent trends. This has proved to improve the accreditation process efficiency in comparison to a paper-based procedure and along with enhancement in the faculty participation. However, it could not demonstrate the statistical comparison between paper-based manual processes and ACAT-based accreditation system (Sabir et al., 2018).

ACAT (ABET Course Assessment Tool) was developed at the computer science and engineering department of the University of Nevada, Reno for streamlining and automating the existing manual course assessment process. ACAT is a web-based application specifically designed for university faculties that collect the assessment data related to the direct and indirect assessment instruments (AIs). ACAT was reportedly designed to address the ABET criteria 3 which is concerned with SO assessment. The direct assessment instruments include data related to each course such as student grades in assignments, quizzes, and examinations whereas indirect assessment instruments include student course evaluation surveys, job placement statistics, and alumni surveys. This allows creating the custom COs apart from what is provided by ABET in criteria 3. The standard assessment report can be generated from this tool based on entered data and accreditation requirements. ACAT uses dynamically generated HTML to display pages to the user with the help of server-side scripting. ACAT uses MySQL database as it offers a stable and free solution. The entire ACAT solution architecture consists of three layers such as presentation layer, business logic layer, and utility layer. The presentation layer mainly includes a GUI. The business logic layer is responsible for implementing various functionalities. The utility layer includes HTML, PHP, and MySQL subsystem. A usability study, aimed at the determination of whether the ACAT tool meets the design goals and how much it is valuable to target customers was performed for ascertaining the effectiveness of this tool. A computer system usability questionnaire was employed to evaluate the ACAT system's usefulness, interface quality, information quality, and overall satisfaction. The system usefulness criteria received the highest rating of 79.7% indicating successful usability of this tool. However, this survey indicated that there was a scope of improvement for "information quality", criteria. Furthermore, the authors reported that this tool is efficient in streamlining the ACAT accreditation process for an engineering program but no evidence whatsoever was presented as to much manual efforts were eased by using this tool (Eugene Essa, 2010).

IonCUDOS is a commercial web-based application available to manage the NBA accreditation process. It provides the complete solution for activities essential in an accreditation cycle such as collection, organization, and analysis of data with a facility to generate reports for fulfilling various criteria related to the program, curriculum, SO assessment, faculty information, and other details required to prepare the SAR. It is also possible to monitor the individual student's performance for a specific course as well as at the overall program level by using outcome attainment reports generated by IonCUDOS for every individual student. This type of handy data related to individual student performance can be well utilized for remedial actions plan which is an essential part of continuous improvement strategy as a part of criteria 7 of NBA (Mahadevaiah et al., 2018; Sawant, 2017). The availability of features such as curriculum design (PO, PSO, PEO, etc.), course design, its delivery, and assessment (online/offline, blended, internals, mid-semester examinations, term-end examinations, seminars, rubrics, weightages, etc.) keeping three domains of knowledge i.e., cognitive, affective, and psychomotor along with bloom's taxonomy in mind makes it a very good tool for the utilization on the path of preparing for accreditation. Furthermore, managing lesson schedule with the help of topics and teaching–learning objectives is easy along with analysis of continuous internal evaluation and end semester question papers which are important as a part of self-audit. The possibility of customization for evaluation of students to get direct attainment via calculating various academic outcomes is another feature that is beneficial to the academic fraternity. The calculations may be based on threshold-based, normal average, or weighted average for CO, CO-PO, and CO-PSO attainments. Also, the software can very well be employed for indirect attainments, such as planning and conducting surveys of various stakeholders including students, employers, alumni, parents, etc., and combine both the attainments to get to the final attainment values along with the provision of giving the action plan to improve the numbers in forthcoming years.

The institutional Enterprise Resource Planning (ERP) systems are more authentic in terms of collective and holistic collection and maintenance of data due to integration of various modules in a single platform such as admissions, finance, class management, faculty management, course/program attainment, etc., e.g., the number of students enrolled for a specific program will be directly reflected through the admission module and continue till the program's attainment (Hart & Snaddon, 2014). The role and effectiveness of ERP systems are scarce in terms of engineering education but have certainly picked up in recent years. These customizable ERP systems are well suited to address the organizational requirements in terms of administrative management. However, they are reported to be poor at academic management and accreditation requirements (Abdel-Haq, 2020).

9.8 CONCLUSION

The accreditation is considered a must in higher education in engineering. It can be looked at from various aspects i.e., declaration of quality practices at the institute which leads to improving rates of enrollment, societal recognition, industrial recognition in terms of credibility and skill set of outgoing students. It also opens the door for international mobility of engineering graduates. The national governments are keen to push their HEIs toward pursuing quality management practices along with getting involved in continuous quality monitoring and improvement to get global recognition. These processes involve a lot of planning, implementation, and quantification of a humongous amount of data that needs to be collected, recorded, and processed in order to get the desired output measurables/indicators. The automation of such tasks in engineering education requires specialized tools and technologies. As none of the current literature gives a one stop list/review of the tools that can be used, the ones discussed herein to aid the preparations for ABET, NAAC, and NBA based SARs and subsequent detailed ones fulfill the contemporary requirements at this current point in time. The time saved on the faculty/administrator's part can be implemented into various academic and research-oriented activities which can again be fruitful for scoring a number of valuable points in the accreditation process.

REFERENCES

Abdel-Haq, M. S. (2020). Conceptual framework for developing an ERP module for quality management and academic accreditation at higher education institutions: The case of Saudi Arabia. *International Journal of Advanced Computer Science and Applications*, *11*(2), 144–152.

Accreditation Criteria & Supporting Documents | ABET. (n.d.). Retrieved February 23, 2021, from https://www.abet.org/accreditation/accreditation-criteria/

Alexander, B., Ashford-Rowe, K., Barajas-Murph, N., Dobbin, G., Knott, J., McCormack, M., Pomerantz, J., Seilhamer, R., & Weber, N. (2019). *Horizon report 2019 higher education edition*. EDU19. Retrieved June 5, 2021 from https://www.learntechlib.org/p/208644/

Anand, S. S., Devaraj, A. F. S., Devi, R. K., Subramanian, C. B., Subramanian, R. R., & Nagaraj, P. (2021). Effective design and implementation of B. Tech (CSE) curriculum with industry tie-ups. *Journal of Engineering Education Transformations*, *34*, 191–200.

Ates, H., & Alsal, K. (2012). The importance of lifelong learning has been increasing. *Procedia – Social and Behavioral Sciences*, *46*, 4092–4096. https://doi.org/10.1016/j.sbspro.2012.06.205

Balasangameshwara, J. (2015). Uncovering the value of ICT in time management for implementation of OBE courses. *Journal of Engineering Education Transformations*, *29*(1), 43–54.

Balikaeva, M. B., Chizhevskaya, E. L., Grevtseva, G. Y., Kotlyarova, I. O., & Volkova, M. A. (2018). Innovative technologies as a means of the development of future engineers' professional mobility abroad. *IOP Conference Series: Materials Science and Engineering*, *441*, 012007. https://doi.org/10.1088/1757-899X/441/1/012007

Banthiya N. K. (2012). *Accreditation of undergraduate engineering degree programmes in india: Changes due to Washington Accord. Journal of Engineering, Science & Management Education*, 5(2), 445–451.

Bovill, C., & Woolmer, C. (2019). How conceptualisations of curriculum in higher education influence student-staff co-creation in and of the curriculum. *Higher Education*, *78*(3), 407–422.

Bruçaj, S. (2019). Quality at newly established private universities: New strategies for leadership management. *European Journal of Education*, *2*(1), 16–21. https://doi.org/10.26417/ejed-2019.v2i1-49

Cavus, N., & Zabadi, T. (2014). A comparison of open source learning management systems. *Procedia-Social and Behavioral Sciences*, *143*, 521–526.

Changwong, K., Sukkamart, A., & Sisan, B. (2018). Critical thinking skill development: Analysis of a new learning management model for Thai high schools. *Journal of International Studies*, *11*(2) 37–48.

Cheong Cheng, Y., & Ming Tam, W. (1997). Multi-models of quality in education. *Quality Assurance in Education*, *5*(1), 22–31. https://doi.org/10.1108/09684889710156558

Cullen, J., Joyce, J., Hassall, T., & Broadbent, M. (2003). Quality in higher education: From monitoring to management. *Quality Assurance in Education*, 11(1), 5–14.

Cuthbert, R. (2010). Students as customers. *Higher Education Review*, *42*(3), 3–25.

Davies, J., Hides, M. T., & Casey, S. (2001). Leadership in higher education. *Total Quality Management*, *12*(7–8), 1025–1030. https://doi.org/10.1080/09544120120096197

Dew, S. K., Lavoie, M., & Snelgrove, A. (2011). *An engineering accreditation management system. Proceedings of the Canadian Engineering Education Association (CEEA).*

Dİşlen, G., Ve, Ö., İfadeleri, Ö., Motivasyon, İ., Sosyal, A., & Dergisi, A. (2013). *The Reasons of Lack of Motivation from the Students' and Teachers' Voices. The Journal of Academic Social Sciences*, *1*, 35–45.

Elhassan, A. S., & El-Hassan, W.-S. (2019). *Assessment Instruments for Accreditation: A Data Management System Design & Implementation. International Journal of Advanced Trends in Computer Science and Engineering*, *8*(1.1), 1–7.

Essa, E. (2010). *ACAT: ABET Course Assessment Tool* [Thesis]. https://scholarworks.unr.edu//handle/11714/4307

Essa, E., Dittrich, A., & Dascalu, S. (2010). *ACAT: A web-based software tool to facilitate course assessment for ABET accreditation. 2010 Seventh International Conference on Information Technology: New Generations*, 88–93. https://doi.org/10.1109/ITNG.2010.224

Ghobakhloo, M., & Fathi, M. (2019). Corporate survival in Industry 4.0 era: The enabling role of lean-digitized manufacturing. *Journal of Manufacturing Technology Management*.

Guilbault, M. (2016). Students as customers in higher education: Reframing the debate. *Journal of Marketing for Higher Education*, *26*(2), 132–142.

Hart, C. A., & Snaddon, D. R. (2014). The organisational performance impact of ERP systems on selected companies. *South African Journal of Industrial Engineering*, *25*(1), 14–28.

Hay, M. (2008). Business schools: A new sense of purpose. *Journal of Management Development*, *27*(4), 371–378. https://doi.org/10.1108/02621710810866723

http://www.naac.gov.in/. (2021). http://www.naac.gov.in/

https://ehea.info/. (2021). https://ehea.info/

https://www.Ieagreements.org/. (2021). https://www.ieagreements.org/

Hussain, W., Spady, W. G., Naqash, M. T., Khan, S. Z., Khawaja, B. A., & Conner, L. (2020). ABET accreditation during and after COVID19—Navigating the digital age. *IEEE Access*, *8*, 218997–219046. https://doi.org/10.1109/ACCESS.2020.3041736

Ibrahim, W., Atif, Y., Shuaib, K., & Sampson, D. (2015). A web-based course assessment tool with direct mapping to student outcomes. *Educational Technology and Society*, *18*(2), 46–59.

Ismail, M. A. (2016). OBACIS: Outcome based analytics and continuous improvement system. *Proceedings of the Canadian Engineering Education Association (CEEA)*. https://doi.org/10.24908/pceea.v0i0.6494

Ismail, M. A. (2017). OBACIS phase III: Accreditation and grading sheets (AGSs) — The excel-app. *Proceedings of the Canadian Engineering Education Association (CEEA)*. https://doi.org/10.24908/pceea.v0i0.10593

Jagadeesh, R. (2000). Assuring quality in management education: The Indian context. *Quality Assurance in Education.*

Jha, S., & Kumar, M. (2012). Management education in India: Issues & challenges. *Available at SSRN 2140807.*

Kasuba, R., & Ziliukas, P. (2004). A comparative review of two major international accrediting consortia for engineering education: The Washington accord and the bologna process. *World Transactions on Engineering and Technology Education*, *3*(1), 71–74.

Kaupp, J., & Frank, B. (2014). *Evaluation of software tools supporting outcomes-based continuous program improvement processes: Part 2. Proceedings of the Canadian Engineering Education Association (CEEA).*

Kettis, Å., Ring, L., Gustavsson, M., & Wallman, A. (2013). Placements: An underused vehicle for quality enhancement in higher education? *Quality in Higher Education*, *19*(1), 28–40. https://doi.org/10.1080/13538322.2013.772697

Komives, C. (2015). Towards quality and consistency in Indian engineering education. *Journal of Engineering Education Transformations*, *29*(1), 1. https://doi.org/10.16920/jeet/2015/v29i1/77100

Kubota, R. (2017). Globalization and language Education in Japan. In N. Van Deusen-Scholl & S. May (Eds.), *Second and Foreign Language Education* (pp. 287–299). Springer International Publishing. https://doi.org/10.1007/978-3-319-02246-8_24

Kumar, S., & Dash, M. K. (2011). Management education in India: Trends, issues and implications. *Research Journal of International Studies*, *18*(1), 16–26.

Lu, J., Laux, C., & Antony, J. (2017). Lean Six Sigma leadership in higher education institutions. *International Journal of Productivity and Performance Management*, *66*(5), 638–650. https://doi.org/10.1108/IJPPM-09-2016-0195

Lynch, H. F., Abdirisak, M., Bogia, M., & Clapp, J. (2020). Evaluating the quality of research ethics review and oversight: A systematic analysis of quality assessment instruments. *AJOB Empirical Bioethics*, *11*(4), 208–222. https://doi.org/10.1080/23294515.2020.1798563

Mahadevaiah, Y., Joshi, M., & Bewoor, A. (2018). *Continual improvement of overall performance to achieve academic excellence in education by automating accreditation processes using IonCUDOS.* Center for Educational Society.

Mallikarjuna, B., Dayananda, P., Niranjanamurthy, M., Kumar, P., & Sabharwal, M. (2020). Outcome-based education for computer science in e-learning through moodle. *PalArch's Journal of Archaeology of Egypt/Egyptology*, *17*(9), 8740–8752.

Martin, M., & Stella, A. (2007). *External quality assurance in higher education: Making choices. fundamentals of educational planning.* ERIC.

Matters, W. D. L. (2016). *Higher education quality: Why documenting learning matters.* Urbana, IL: University of Illinois and Indiana University, Author.

Mikhnenko, G., & Absaliamova, Y. (2018). The formation of intellectual mobility of engineering students through integration of foreign language education and professional training. *Advanced Education*, *5*, 33–38.

Mohammed, A. H., Taib, C. A. B., & Nadarajan, S. (2016). Quality management practices, organizational learning, organizational culture, and organizational performance in Iraqi higher education institution: An instrument design. *International Journal of Applied Business and Economic Research*, *14*(14), 1–19.

Namoun, A., Taleb, A., & Benaida, M. (2018). An expert comparison of accreditation support tools for the undergraduate computing programs. *International Journal of Advanced Computer Science and Applications*, *9*. https://doi.org/10.14569/IJACSA.2018.090948

Natarajan, R. (2000). The role of accreditation in promoting quality assurance of technical education. *International Journal of Engineering Education*, *16*(2), 85–96.

Nikolantonakis, K. (2006). *Weights and measures: The Greek efforts to integrate the metric system.* 456–462. http://www.2iceshs.cyfronet.pl/2ICESHS_Proceedings/Chapter_16/R-8_Nikolantonakis.pdf

Obermiller, C., Fleenor, P., & Peter, R. (2005). Students as customers or products: Perceptions and preferences of faculty and students. *Marketing Education Review*, *15*(2), 27–36.

Osipian, A. L. (2014). Will bribery and fraud converge? Comparative corruption in higher education in Russia and the USA. *Compare: A Journal of Comparative and International Education*, *44*(2), 252–273.

Oza, V., & Parab, S. (2011). Quality management education in India in the 21st Century 7-14-21 Model. *International Conference on Advancements in Information Technology with Workshop of ICBMG*, Available at: www.Ipcsit.Com/Vol20/44-ICAIT2011-G1018. Pdf (Accessed 1 July 2012).

Pavel, A.-P. (2012). The importance of quality in higher education in an increasingly knowledge-driven society. *International Journal of Academic Research in Accounting Finance and Management Sciences*, *2*(Special 1), 120–127.

Prados, J. W., Peterson, G. D., & Lattuca, L. R. (2005). Quality assurance of engineering education through accreditation: The impact of engineering criteria 2000 and its global influence. *Journal of Engineering Education*, *94*(1), 165–184. https://doi.org/10.1002/j.2168-9830.2005.tb00836.x

Pratibha, D., Anurag, P., Nagamani, C., & Keerthi, D. S. (2021). Impact of industry collaboration in developing core engineering departments. *Journal of Engineering Education Transformations*, *34*, 468–476.

Prisacariu, A., & Shah, M. (2016). Defining the quality of higher education around ethics and moral values. *Quality in Higher Education*, *22*(2), 152–166. https://doi.org/10.1080/13538322.2016.1201931

Quinn, A., Lemay, G., Larsen, P., & Johnson, D. M. (2009). Service quality in higher education. *Total Quality Management*, *20*(2), 139–152.

Rao, P. S., & Hans, K. (2011). Comparative analysis of accreditation systems in management education in India (NBA AND SAQS). *Aweshkar Research Journal*, *12*(2).

Romero, M., Lepage, A., & Lille, B. (2017). Computational thinking development through creative programming in higher education. *International Journal of Educational Technology in Higher Education*, *14*(1), 1–15.

Sabir, A., Abbasi, N. A., & Islam, M. N. (2018). An electronic data management and analysis application for ABET accreditation. *ArXiv:1901.05845 [Physics]*. http://arxiv.org/abs/1901.05845

Sahay, B. S., & Thakur, R. R. (2007). Excellence through accreditation in Indian B-Schools. *Global Journal of Flexible Systems Management*, *8*(4), 9–16.

Sawant, P. (2017). Course outcomes attainment analysis using automated tool-IONCUDOS. *Journal of Engineering Education Transformations*, Special Issue, eISSN 2394-1707.

Schahczenski, C., & Van Dyne, M. (2019, January 8). *Easing the Burden of Program Assessment: Web-based Tool Facilitates Measuring Student Outcomes for ABET Accreditation*. https://doi.org/10.24251/HICSS.2019.919

Shankar, R., Dickson, J., & Mazoleny, C. A. (2013). *A tool for ABET accreditation. 120th American Society Engineering Education Conference & Exposition.*

Shatunova, O., Anisimova, T., Sabirova, F., & Kalimullina, O. (2019). STEAM as an Innovative Educational Technology. *Journal of Social Studies Education Research*, *10*(2), 131–144.

Shiramizu, S., & Singh, A. (2007). Leadership to improve quality within an organization. *Leadership and Management in Engineering*, *7*(4), 129–140. https://doi.org/10.1061/(ASCE)1532-6748(2007)7:4(129)

Si, J. (2019). English as a native language, World Englishes and English as a lingua franca-informed materials: Acceptance, perceptions and attitudes of Chinese English learners. *Asian Englishes*, *21*(2), 190–206. https://doi.org/10.1080/13488678.2018.1544700

Sinha, V., & Subramanian, K. S. (2013). *Accreditation in India: Path of Achieving Educational Excellence* (SSRN Scholarly Paper ID 2239206). Social Science Research Network. https://papers.ssrn.com/abstract=2239206

Statistica Inc. (2019, September). *Number of higher educational institutions across India in 2019, by type*. Statista. https://www.statista.com/statistics/660862/higher-education-institutions-bytype-india/

Stella, A. (2002). *External Quality Assurance in Indian Higher Education: Case Study of the National Assessment and Accreditation Council (NAAC). New Trends in Higher Education*. International Institute for Education Planning, 7–9 rue Eugene Delacroix, 75116 Paris, France.

Sthapak B. K. (2012). Globalisation of Indian engineering education through the Washington Accord. *Journal of Engineering, Science & Management Education 5*(2), 464–466.

Sun, S., Qi, Y., & Wu, P. (2017). *The Design and Implementation of an Engineering Education Accreditation Aided Management System*. 82–85. https://doi.org/10.2991/icsshe-17.2017.21

Teeroovengadum, V., Kamalanabhan, T. J., & Seebaluck, A. K. (2016). Measuring service quality in higher education. *Quality Assurance in Education*, *24*(2), 244–258.

UGC. (2017). *Growth of higher education in India*. University Grants Commission. https://www.ugc.ac.in/stats.aspx

UGC. (2020). *Consolidated list of all universities in India* (p. 41). https://www.ugc.ac.in/oldpdf/consolidated%20list%20of%20All%20universities.pdf

10 4QS Predictive Model Based on Machine Learning for Continuous Student Learning Assessment

Sunny Nanade
NMIMS University, India

Sachin Lal
Sir Padampat Singhania University, India

Archana Nanade
NMIMS University, India

CONTENTS

10.1 INTRODUCTION

Throughout the twenty-first century, educators face daunting obstacles. They are intended to educate today's youth for employment in the multinational industries that comprise of global society, using an education system that was created over a

DOI: 10.1201/9781003102298-10

century ago to train factory employees in foreign countries to follow orders in production line jobs.

Today's educators are required to include a laundry list of gradual results following the same initial method in addition to providing reading, writing, and arithmetic lessons to school-age children as they were a century and a half ago. All deliverables have basic curriculum capabilities, often without proper facilities to accomplish the mission, pre-and post-school daycares, childhood, and family planning services, extended education opportunities, therapy tools, ethnic preparation, post-secondary training, recreation, and sports events (and also equitably for all genders). Furthermore, prospective immigrants with special education programs that are not or are inadequately provided by the federal government, as well as English as second language services, must be provided by them. Within one year, these language programs are required to solve all obstacles for future English language learners, allowing them to succeed at the same standard of government-mandated competence in all subject areas as all students whose primary language is English. Our teachers are expected to provide transportation to and from school and community education facilities for adult students. They must be accredited to keep their credentials up to date by participating in continuing professional development classes in the area for which they are responsible. Discretionary, state, and local authorities keep them accountable for both student academic success and advancement within a process that must be decided upon before allocation of federal funds. Our once-simple public education system has evolved into a complicated hierarchical monolith that is often beyond our educators' influence.

The growth of production that is lean, design that is lean and lean is major and biggest advancement of industrial preaching over the two decades of the past (Alsmadi, Almani, & Jerisat, 2012). In 1990, Womack and Jones invented the term "light" which was used to characterize the System of Toyota Development, which has been capable of producing more output with minimum energy. Development using lean's method targets waste reduction and taking care that the stages that are remaining run (Womack & Jones, 1996). The technique also contributes to significantly higher outcomes when conducted correctly. Although lean was evolved and established in the industry of manufacturing, it has in present times been used with a little degree of service performance (Alsmadi et al. 2012).

The time has come for those born in this millennium to enroll in universities. It is necessary to rethink and redefine the way the education system should be changed. The curriculum and the system followed by the institutes are being reconstructed to various questions to alter the educational system. The older system focused on what is taught rather than to whom it was taught or the method of teaching, and this system is getting outdated. Due to the rise of technological advancement, the learning style is being altered by the students to internet-based learning along with machine learning to predict the solution since the last decade. This has created problems since the educational institutes haven't kept up with the changing technology. Students who are good with technology have already come up with new ways to learn through the internet. This makes their curriculum not functional enough (Hines et al. 2008). The students have connections with the other learners for collaborative learning and can teach each other the concepts they don't understand. The educators have to understand the upcoming technology. They also have to be more familiar with them to make improvements in the teaching methodology with the predictive method to improve the performance of the system. When more innovation is made in the curriculum using ML, the students will have more scope to solve problems and it will become more interactive. These new methodologies will be able to keep up with the learners and will change and develop as quickly as they (Crute et al. 2003).

Lean education is the implementation of lean principles in the field of education using the predictive method. The lean principle is the practice of improving something continuously. The principles have to be followed in the educational system by frequently adapting the curriculum to the changing world. This adaptation in the syllabus, infrastructure, and methodology is difficult for schools and universities. Usually, the curriculum is changed every four years, but changing it faster than that costs a lot to the university. It is also difficult to keep with the quickly changing world. Various innovations take place during this period making the syllabus absolute and outdated. However, if the educators add the new technology immediately when it is still in the development stage, there are chances for the technology to fail. Hence, the educators must be dynamic and quick learning. They have to adapt to the ever-changing technology to frame the curriculum. However, the learning phase of the educator must be a step ahead of the learners or it will lead to a waste of resources (Womack and Jones 2003).

The waste dilemma is first discussed to help explain lean. A study of principles and strategies which are used to minimize waste is accompanied by this. Subsequently, the problems that need to be taken into consideration in integrating lean facilities are undertaken.

- **Lean in Education**

Lean, which has traditionally been used in the manufacturing sector and is now widely applied in the non-industrial industry, focuses on improving behavior and coherence. Being a waste-reducing technique, demonstration of its effectiveness in the processes of educational development, as well as its advancements in other industrial sector output processes, has begun to be demonstrated in lean. In this way, the field of education is young and Lean methods have proven that they can produce added value by retaining high-level expertise, decreasing school costs, reducing time to plan, saving money, and allowing ready-made procedures to begin (Womack & Jones, 1996). The primary objective as a tool of organizational expansion and change is to increase performance. The challenges that have been faced are not only called a challenge but an ability to turn. For problem-solving, considerations like openness, engagement, speed, and cooperative learning processes are very critical (Dennis, 2007). Lean curriculum administration is a readily implementable paradigm of reform in all schools (Cleary & Duncan, 1997). The greatest obstacle faced by Lean's transition to school is the destructive patterns of students.

To phrase it another way: resisting change and creativity. Training is a framework that encompasses all the knowledge-distribution systems involved. All these systems will be improved by evaluating our educational institutions' systems of management, instruction, and learning. Effective management and teamwork of the workers will achieve lasting performance. It will treat schools, students, families, and teachers in the most successful manner (Scott, 2008). The cornerstone of Lean Education is the methodology and standardization of the college curriculum. Leveling of load, waste reduction, PDCA (Plan- Do-Check-Act) demonstrated in the figure below, parallel research, strategic reasoning, number ten, value addition, philosophy of learning, quality management, brain science, preparation, judgment matrix, a quest for excellence, value flow, authorization, error avoidance, replication, style of teaching, actual scheduling, mapping methods, checkpoint, etc., are the simple components of Lean that can be used (Figure 10.1).

Communication, trust, and commitment are the core themes within the administration–parent–teacher–student quadrant. The growth that is continuous at all Lean stages is a critical one. The roadmap encompasses a comprehensive curriculum and low-cost online learning will build a chain of change beginning with a student and reaching the whole population. Education of lean requires that each individual is involved in a spectrum of institutions of education which recognize and collaborate together in finding new solutions to problems, teaching and learning processes. Usage of Lean approaches to increase performance in the sector of educational institutions will allow educators to put Lean culture into practice (Cleary & Duncan, 2008). This curriculum has all the skills required by the students to gain at the point. The information sum for every step is the material which the person who learns will gain during his / her academic life. However, the curriculum cannot be completed on virtually any level of the academic life of a student due to encountered difficulties. Incorrect curriculum to which students are subjected during their school life allows incomplete information to be created, and gaps in knowledge that cannot be acquired at previous levels prohibit new information from being received. For this purpose, the instruction

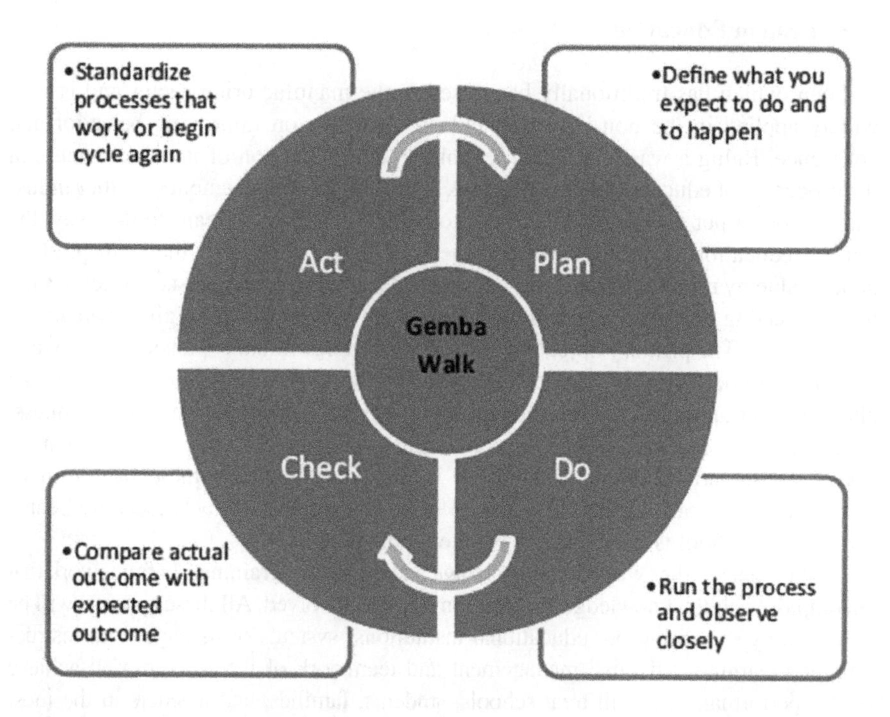

FIGURE 10.1 The Plan-Do-Check-Act (PDCA) cycle.

must be completed. Although Lean ideas, processes, and similar techniques are implemented in the service sector, how those approaches can apply to education has never been conceived. The study explains how to use Lean in the education sector, and how the pupils, parents, teachers, and school's administration quartet can hit the fruitful land of everlasting achievement in the terms that are related to Lean words.

During the research theory of learning, it is stressed that people study faster when they own a model, especially a repeated model. Theory of Learning is based on the fact that each individual's particular fields of intellect and ability to interpret knowledge during the learning process. A person learns in a way that he or she thinks is important. To provide value-added resources for education to all students, an instructor must deliver learning that is centered on students using the best approaches and to provide knowledge. Proper teaching means allowing all learners to take advantage of their learning style while giving different activities with ten times of remind/ recall. Each student will thus have the chance that is best for them to achieve goals of learning. The issue is concerning with the level of learning theory which is high (Jenkins, 2003). The response depends on the precision with which this is done and what time it takes for the teacher to respond to the student. Brain theory depicts the need for an uninterrupted, continuous stream of information that can be easily absorbed by the mind. On the other, the learning philosophy involves several forms of knowledge in ten times the mechanism of repetition to help and inspire our learners as they learn. Apart from this, the bond of trust enables the mechanism to take place.

Lean's definition of waste in education is another aspect of educational adaptation. Lean education's disadvantage is its use in the instructional phase of all elements beyond their capacity, including teachers, pupils, parents, and administrative personnel in the classroom. Our most important commodity in the education of Lean is our human capital and if that is the case, we need to make the most productive use of the time of those individuals. In other terms, the loss of time can be minimized. We define the waste in the education process as follows:

> Failure of students in exams due to unprogrammed training, measurement, and support procedures graduating of students from institutions of education to seek a respectable job and without academic experience of a lifetime, offering classes that students cannot apply in their daily lives. That would not contribute to the sense of significance, program errors, and shortcomings apart from these in the teaching process.

Management of Lean implementation in the education sector also requires the removal of the above-mentioned duplication that has occurred in this phase. Schools that wish to implement a lean approach must first define any vacant spaces that could be included in the schooling spectrum. The following strategies can be used to make schools run effectively and minimize unnecessary waste:

> To organize instructional areas and to put scarce equipment, materials, and files in order, offices and work areas organized, setting a standard to the location of staff, training facilities, and files, clearing areas of work and schools to prevent problems that could result from misunderstandings, pollution, etc., Holding oversaw these principles (Dahlgaard & Stergaard, 2000).

From the review on the various lean educational systems and the simulation process, it has been seen that the researchers have focused mainly on the organizational and the management side of the lean system. The study of lean processes in the education sector is rare and simulations have not been done effectively. From these aspects, the objectives proposed in this chapter are:

- To build lean-based process for engineering curriculum development
- To assess the educational value of the lean-based developed system

10.2 MODELS FOR STUDENT LEARNING ASSESSMENT

a) **Quality Management Models**

Quality management: Methods relating to successful growth and maximizing efficacy. Goods and processes combined formed the Total Quality Management (TQM) kit (Shah & Wall, 2003). Consistency processes usually emphasize the importance of cross-functional software design as well as hierarchical process management. The role of customers, suppliers, and workers in ensuring continuity is also highlighted (Cua, McKone, & Schroeder, 2001).

Lean production involves choosing products from the lean warehouse to attain excellence technically. There is a risk of leaning heavily on the instrument's striving and benefit for the excellence of the process but the elimination of the durability of the lean approach within the particular work community. When the latest strategies are adopted, there is an organizational risk: both a hard and an opportunity. Relevant issues are:

(1) What are the consequences under discussion of the lean solution? How difficult or how likely they are to be achieved? (Does that make doing this worth it?)
(2) How does the use of the lean approach lead to the company's sustainability in the transformation and its benefits? (Are there long-term savings from doing so?)

We are primarily interested in the operational applicability of lean resources in this article, and in business decision-making that precedes, in particular, lean implementation.

There are three questions: which of the various lean instruments should be used in a given situation; how the risks (opportunities and threats) of each candidate instrument can be identified; and how corporate culture is affected by the success (or otherwise) of applying change management (Figure 10.2).

There are three issues: which of the different lean approaches to be used in a given context; how to deal with the problems (opportunities and threats) with each candidate tool, and how the effectiveness (or otherwise) of implementing change management impacts organizational conduct; how to use lean to strengthen the education system.

b) **Other models used for the teaching–learning process**

In assessing productive learning methods based on the evolving interests of learners by the creation of new skills, instructors and clinicians adopt the quotient model of

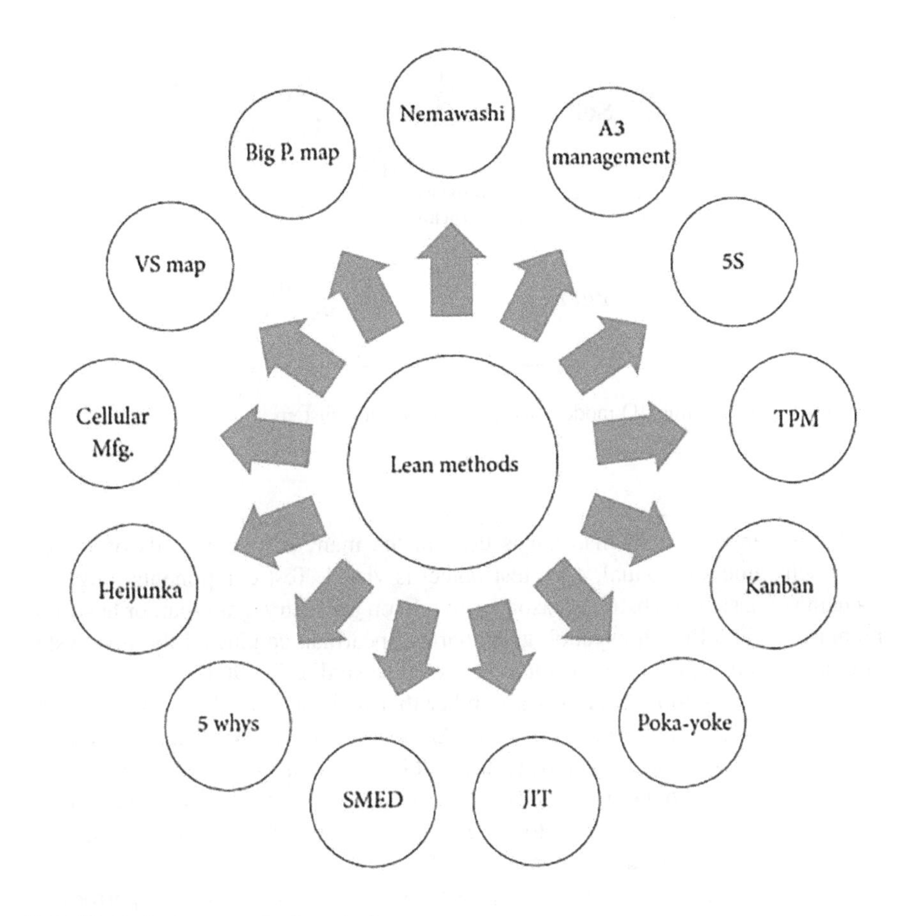

FIGURE 10.2 Lean methods or tools.

assisted learning. Since competence preparation is adopted as a coping mechanism for occupational therapists, it provides a way to understand, organize, and manage the execution of learning strategies. The aim is to increase efficiency through the manufacture of workplace output components in the target occupation.

The first effort to establish a standardized metric for rating the intellect of a person was made by the English statistician Francis Galton. He claimed that intelligence was a result of heredity ("which he did not mean evolution, although he developed some pre-Mendelian theories of particulate inheritance"), as a pioneer in psychometrics and the application of statistical techniques to the study of human growth and the study of the inheritance of human characteristics. He suggested that intelligence and other physical traits like reflexes, body strength, and head size could be correlated. He founded the first mental training facilities in the world in 1882 and published "Inquiries into Human Faculty and Its Development" in 1883, in which he set out his theories. After gathering data on a variety of physical variables, he was unable to show any such correlation, and he eventually abandoned this research.

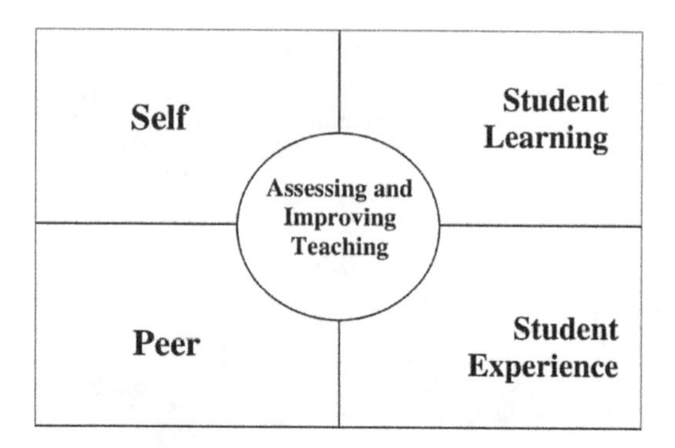

FIGURE 10.3 Exiting 4Q model used by the education system.

A vast range of item material is used in the many different kinds of IQ tests. Although some are textual, each test object is visual. Test components range from focusing on issues of abstract reasoning to rely on geometry, grammar, or basic comprehension. The British psychologist Charles Spearman conducted the first systematic factor analysis of the connectors between the studies in 1904.

An adversity quotient (AQ) is a number that indicates the ability of a person to cope with adversity in his or her life. The AQ also influences the understandability of the pupil's mathematics, as per W. Hidayat. It is widely called the study of durability. In his book *Adversity Quotient: Converting Challenges into Rewards*, the term was coined by Paul Stoltz in 1997. Stoltz developed an estimation tool called the adversity reaction profile (ARP) to measure the AQ.

The AQ (Figure 10.3) is one of the possible markers of an individual's performance in life and is also most helpful in a forecasting mood, mental tension, perseverance, longevity, understanding, and response to changes in the environment.

The paradigm of Schwandt (2009) has a beneficial synergy with our layout and offers a valuable macro-level first port of call to understand your approach to analyzing the FUT software. Similar to the 4Q model stated by Smith (2008), we chose another way of conceptualizing evidence. For multiple reasons, this alternate method helps us to construct an image of facts.

10.3 ROLE OF MACHINE LEARNING IN PREDICTION

"Machine learning is defined as a computer science discipline that uses mathematical techniques to give computer systems the ability to 'learn' from data without being specifically programmed for readers who are new to the word (i.e., constantly increasing performance on a given task)." By using Machine Learning, one can improve the overall performance of a learner by predicting the results based on input parameters such as his SSC, HSC, 4QS marks. By mapping 4QS test marks with Bloom's taxonomy, one can design the curriculum best suited to the learner that would reduce educational waste.

How can machine learning revolutionize the field of education?

I. **By Increasing the Efficiency**: By integrating activities like classroom management, scheduling, etc., machine learning can make educators more effective in artificial intelligence.
II. **Learning Analytics**: When it comes to learning analytics, machine learning can help teachers understand better the information that cannot be gleaned by the usage of the human brain.
III. **By Predictive Analytics**: When it comes to predictive analytics, machine learning could draw answers about situations that may occur in the future.
IV. **By Adaptive Learning**: Adaptive learning is a technology-driven or online instructional framework that analyzes the success of a pupil in real time, and based on the input, it changes teaching methods and the curriculum. Machine learning could be used to remediate failing pupils or challenge gifted individuals in the form of personalized learning.
V. **Personalized Learning**: Personalized learning is an instructional style in which learners direct their learning, go at their speed, and make their own choices on what to learn in certain instances. Machine Learning in the form of personalized learning may be used to give each pupil an individualized educational experience.
VI. **Assessment**: Using the Machine Learning in the education sector, we can predict the students' performance on their histological dataset to focus on their future to improve their stability with their growth. AI can be used to grade student assignments and reviews accurately.
   a) **K-NN Algorithm for Machine Learning**
      i. Known as one of THE simplest Machine Learning Algorithms K-Nearest Neighbor is based on the Supervised Learning technique. This algorithm assumes the similarity between the new case/data and available cases and puts new instances into the category that is most similar to the available categories (Figure 10.4). This means when new data appears then it can be easily classified into a good suite category by using the K-NN algorithm. This algorithm is used for classification problems. K-NN a non-parametric algorithm which means it doesn't make any assumptions on underlying data.

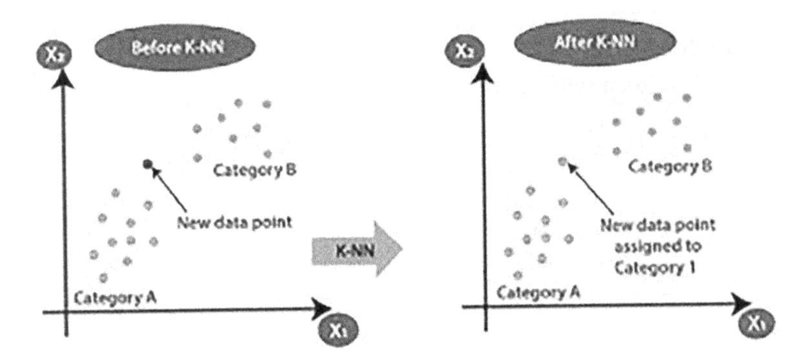

FIGURE 10.4 K-NN Algorithm for Machine Learning.

b) **Random Forest Algorithm for Machine Learning**

   i. Based on Ensemble Learning, Random Forest is a popular machine learning algorithm that belongs to the supervised learning technique. "Random Forest is a classifier that contains several decision trees on various subsets of the given dataset and takes the average to improve the predictive accuracy of that dataset (Figure 10.5)."

c) **Test and exam result prediction**

   The following section gives a detailed overview of education and curriculum development on different Quotients (Table 10.1):

   - Written tests were conducted based on the QS model to test the quotients of students at the beginning.
   - Questioners were shared with college students in the vicinity.
   - Data collected includes deemed university, government-aided college, and minority college.

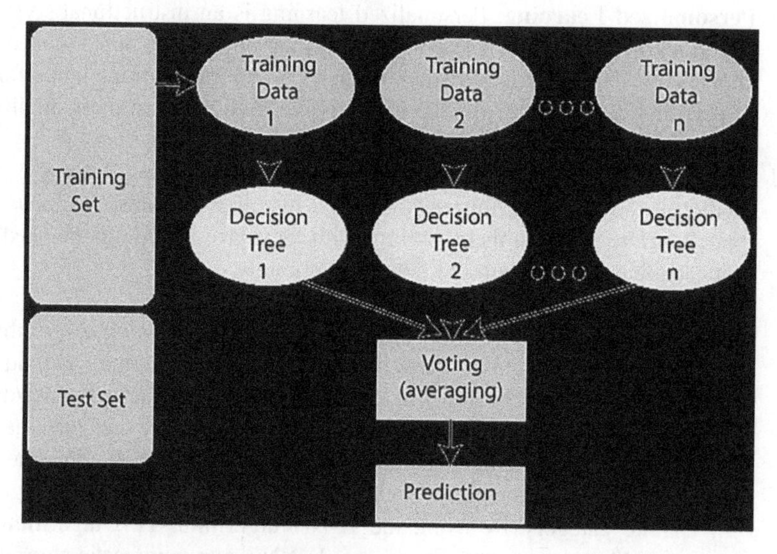

FIGURE 10.5 Random Forest Algorithm for Machine Learning.

TABLE 10.1
Test pattern with their details

Sr. No.	Quotient Tested	Number of Questions	Time Duration
Q1	Adversity Quotient	20	30 mins
Q2	Intelligence Quotient	30	60 mins
Q3	Emotional Quotient	15	15 mins
Q4	Spiritual Quotient	15	15 mins

d) **Test conduction on some sample dataset (Table 10.2)**

TABLE 10.2
Different branch college students collected data

College	Branch	Intake	Responses	Final	Total
C1	CMPN	60	59	58	
C1	IT	60	59	55	
C1	MECH	60	58	57	215
C1	MTRX	60	45	45	
C2	CHEM	60	30	28	
C2	CIVIL	60	47	45	129
C2	MECH	60	57	56	
C3	CMPN	60	52	50	
C3	IT	60	57	49	99
C4	CIVIL	120	52	31	
C4	CMPN	180	49	38	114
C4	MECH	120	46	45	

10.4 4QS PREDICTIVE MODEL

From the literature review on the various lean educational systems and the simulation process, it is seen that the researchers have focused mainly on the organizational and the management side of the lean system. The study of lean process in the education sector is very rare and simulations have also not been done effectively.

From these aspects, the objectives of the proposed research work are as follows:

1. Implement the 4QS model.
2. Map the 4QS model with Cognitive Levels.
3. To assess the educational value of the lean-based developed system into the framework. From the literature review, various gaps have been found and they have been given in the objectives as the steps for this project.

To make the study helpful, the Model Syllabus as proposed by AICTE for Engineering will be studied.

Steps 1 and 2 mentioned above can be by taking the 4QS test and mapping the test results as shown in Figures 10.6 and 10.7.

The test result of the 4QS test is mapped with the respective learning level and the same is tabulated for the entire class. Class-wise data for each test broken down into three categories say High (H), Medium (M), and Low (L). The instructor can now design the respective COs based on the level of understanding of the class.

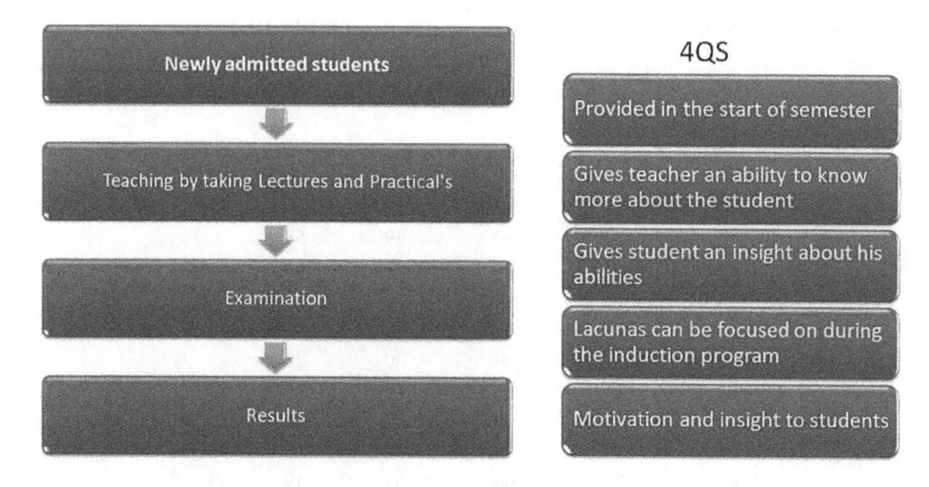

FIGURE 10.6 4QS Model used in the education system.

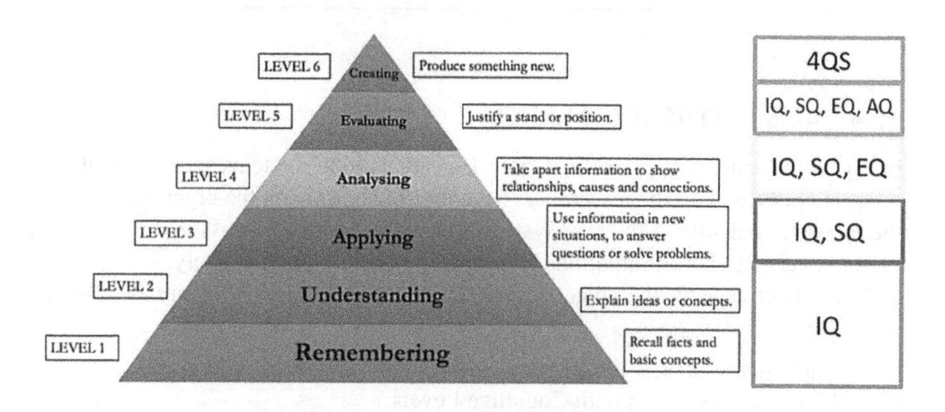

FIGURE 10.7 Mapping of 4QS test with the cognitive domain.

10.5 4QS MODEL FOR IMPROVING LEARNING EDUCATION

The following study presents how 4QS Lean, a technique traditionally used only in education, is tailored to the particularities of the higher education processes and is implemented in a college/university. The four-quadrant model of eased learning (4QS) is used by teachers and practitioners to select effective learning approaches based on learners' changing preferences through the acquisition of new skills. As occupational therapists apply competence acquisition as a rehabilitation strategy, the 4QS supplies a way to define, organize, and implement the execution of therapeutic strategies. The aim is to increase efficiency in the desired occupation through the development of occupational output components. Figure 4 4QS Model used in the education system. Following are the four quotients that shall be accessed with stress analysis using the 4QS model:

1. Intelligence
2. Adversity
3. Spiritual
4. Emotional
5. Stress

1) **Intelligence quotient (IQ)**: IQ as another predictor of student success. In individual creative acts, like aspects of music composition, mathematics, and writing, a high IQ is essential as well as for student performance by using some intelligence questions. And clearly, a high IQ is usually desired for innovators as well for predicting the result. But innovation is a collaborative effort and genius often comes from putting together a team with complementary skills. Here we are going to use some sample questions and ask the student to solve that before starting their first semester with other quotients for predicting their result of the first semester so we can focus more on their weakness.

   Historically, before the IQ tests being invented, there were attempts made to classify people into intelligence groups by examining their daily life behavior. Many other forms of behavioral testing are also relevant for the validation of classifications primarily based on IQ test scores. Both the classification of intelligence by behavioral evaluation in the training room and the classification by IQ tests depend on the word "intelligence" used in the specific case and the mistake in the classification phase in terms of reliability and measurement.

   The first effort to establish a standardized metric for rating the intellect of a person was made by the English statistician Francis Galton. As a leader in the application of statistical techniques and psychometrics to human development research and hereditary analysis of characteristics in humans, he declared that high intellect was predominantly a result of heredity (by that he did not mean evolution, even though he developed some pre-Mendelian particulate heritage theories). The relationship between intellect and other measurable attributes, such as reflexes, muscle power, and head height, was proposed by him. In 1882 he founded the world's first mental learning center, and in 1883 "Inquiries into the Human Faculty and its Growth," was published where he introduced his research findings. He was unable to prove any such correlation after obtaining data on several physical factors and ultimately abandoned this project.

   The various types of IQ tests include a spectrum of materials on products. Any elements of the study are visual and many are textual. Checking stuff ranges from relying on algebra, syntax, or general ability to be focused on logical reasoning concerns (Figures 10.8 and 10.9).

   In 1904 the psychologist Charles Spearman conducted the first systematic factor analysis in association science.

Name:______________________________ **Division:**_____

Test: Q2 **Duration: 60 minutes.**

Circle the correct option

Roll No: ________ **Branch:**______

Date: 09th July 2019

Circle the answer(s), or write in the answer box provided.

1.

50	51	49	52	48
46	47	45	48	44
49	50		51	
47		46		45
48		47	50	46

Which is the missing section?

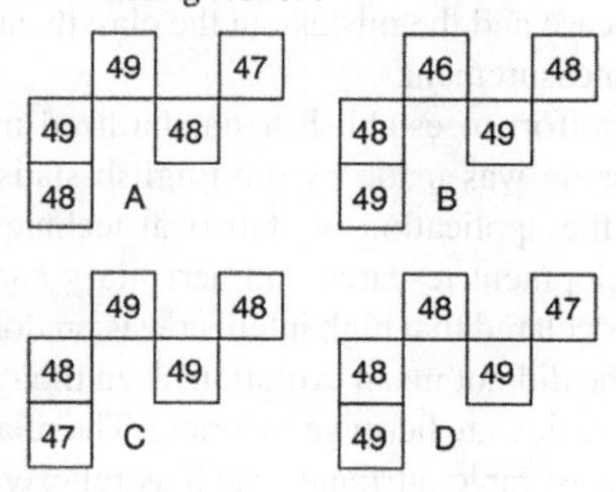

FIGURE 10.8 Sample of Intelligence quotient (IQ).

2) **Adversity quotient (AQ)**: A violent or critical illness is the meaning of the term "adverse." We are met with all of the other adversities throughout life; it is only by adversities that one gets to know who is pleasant and well willed. There is an AQ, which includes SQ (spiritual quotient), EQ (emotional quotient), and IQ (intelligence quotient), which is a score that measures the willingness of an individual over his or her life to deal with adversity. One needs to develop endurance to overcome adversities in life.

Scoring the ARP QuickTake

STEP ONE: Insert the number you circled for each of the above questions into the corresponding boxes below. For example, if you circled 4 for question number 2, you would insert the number "4" next to question number 2 below.

C	O	R	E
1.	2.	3.	4.
7.	6.	5.	8.
13.	11.	9.	10.
15.	16.	12.	14.
17.	18.	20.	19.

STEP TWO: Insert the total for each column in the grid below.

Total C +	Total O +	Total R +	Total E =	Total from CORE

STEP THREE: Add the four totals from Step Two, then multiply that number by 2. Insert your answer in the triangle.

________ x 2 = AQ

Total from CORE

YOUR AQ QUICKTAKE™ SCORE

Control, Ownership, Reach, and Endurance,

AQ Range:

High	AQ (178–200)
Moderately High	AQ (161-177)
Moderate	AQ (135-160)
Moderately Low	AQ (118–134)
Low	AQ (117 and below)

20 Questions
30 Minutes

5 point Likert-type scale
Min and Max criteria's varied depending upon situation given in question

Calculations of Adversity Quotient, **Stoltz (2010)**

FIGURE 10.9 Sample pattern of Intelligence quotient (IQ).

AQ is one of a person's possible success metrics in life and is also often used in assessing mood, mental tensity, perseverance, endurance (Figures 10.10 and 10.11).

3) **Emotional quotient (EQ)**: EQ is your way of managing your thoughts in positive ways so that you can deal with the vicissitudes of life effectively. Studies say that establishing intimate relationships and entering into group environments is better for people with EQ ratings. Individuals with better emotional intelligence (EI) may have a clearer understanding of their psychological state, which may entail treating tensity efficiently and being less likely to experience depression. Emotional Intelligence Quotient (EIQ), Emotional Leadership (EL), Emotional Intelligence (EI), and EQ are people's abilities to interpret, correctly classify, and mark their own and other emotions, use awareness of emotions to guide thoughts and actions, and monitor or alter their sentiments to react to or attain circumstances.

While the word first appeared in 1964, the 1995 book *Emotional Intelligence*, written by the science journalist Daniel Goleman, gained interest.

Empathy is commonly synonymous with EI since it refers to an agency sharing its own emotions with others. There are some models, however, that aim to measure degrees of EI (empathy). Several EI versions are available now. The original Goleman model should be called a hybrid model combining what has been modeled separately as an EI functionality and EI functionality. Goleman described EI as the set of skills and traits that drive leadership effectiveness. The trait pattern was created by Konstantinos V. Petrides in 2001. "It involves patterns of behaviour and self-perceived skill, which

Name: ______________________ **Division:**____ **Branch:**____ **Roll No:**________

Test: Q1 **Duration: 30 minutes.** **Date:** 09th July 2019

Circle the correct option

Imagine the following events as if they were happening right now.
Vividly imagine what will happen as a result of each event (the consequences)
Then circle the number that represents your answer to each of the related questions.

Example:
Situation: You lose your favorite pen. (Imagine this happening to you. Picture in mind)
Imagine what will happen as a result. "I'll never have a pen like that one again. My dad will be so upset when he finds out I lost it."
Circle the number that represents your answer to the questions below each situation.
To what extent can you influence this situation?

Not at all 1 (2) 3 4 5 Completely

---

1. You suffer a financial setback.
   To what extent can you influence this situation?
   Not at all 1 2 3 4 5 Completely
2. You are overlooked for a promotion.
   To what extent do you feel responsible for improving the situation?
   Not at all 1 2 3 4 5 Completely
3. You are criticized for a big project that you just completed.
   The consequences of this situation will:
   Affect all aspects of my life 1 2 3 4 5 Be limited to this situation
4. You accidentally delete an important email.
   The consequences of this situation will:
   Last forever 1 2 3 4 5 Quickly pass

FIGURE 10.10 Sample question of Adversity Quotient (AQ).

30 Questions 1 Marks each
60 Minutes

Scores between: Rating Range of Marks: 0 to 30
27–30 Very highly exceptional
24–26 High expert
21–23 Expert
19–20 Very high average
17–18 High average
13–16 Middle average
10–12 Low average
6–9 Borderline low
3–5 Low
0–2 Very low

FIGURE 10.11 Sample pattern of Adversity Quotient (AQ).

are evaluated through self-reporting." The concept of skill, created in 2004 through Peter Salovey and John Mayer, focuses on the dimensions of the subject to interpret and use emotional knowledge to manage the social world.

We're going to use a sample question here as above and ask students to answer that (see Figures 10.12 and 10.13).

Name: ______________________________ **Division:** _____

Test: Q3 **Duration: 15 minutes.**

Circle the correct option

Roll No: ________ **Branch:** ______

Date: 09th July 2019

Answer each question or statement by choosing which one of the three alternative responses given is most applicable to you.

1. What is your attitude to beggars in the street?
   A I feel somewhat uncomfortable when I see them and, perhaps, think: *there but for the grace of God go I*
   B I feel sorry for them
   C I give them a wide berth

   Answer ☐

2. Do you tend to laugh a lot?
   A About the same as most people
   B More than most people
   C Less than most people

   Answer ☐

FIGURE 10.12 Sample question of Emotional quotient (EQ).

15 Questions
15 Minutes

Total score 25–30 Excessively emotional
Total score 19–24 Fairly emotional
Total score 13–18 Average
Total score 8–12 Fairly unemotional
Total score below 7 Excessively unemotional

1 point for every 'a' answer,
2 points for every 'b',
and 0 points for every 'c'.

FIGURE 10.13 Sample pattern of Emotional quotient (EQ).

4) **Spiritual quotient (SQ)**: SQ goes deeper than your mental and psychological capacity. It's understanding your life and thinking of what you could offer to humanity. It is to live with diffidence; to bear in mind that you are just a tiny one, compared to the vastness of the universe.

The SQ is a parameter that measures the moral acumen of a person; it is as necessary as the intellectual quotient (IQ) and the EQ. Although IQ looks at logical intellect, EQ looks at an individual's emotional ability, and an individual's SQ looks at a person's spiritual strength. Spirituality enhances a person's ability to be imaginative and be receptive and perceptive. You may improve the strength of insight and comprehension, with the assistance of trust. In today's era, why is SQ much more significant? It helps to deal tremendously with the existing problems of terrorism, inconsideration, lack of integrity, and to do away with them. As the next big feature of science studies, the principle of Spiritual Quotient (SQ) increasingly arises. It directly corresponds to the intellect and consciousness of an individual.

Sight (eyes), listening (ears), taste (tongue), nose (scent), and touch (sensational) are the five senses of man. The desire to understand that there is intelligence outside of our five senses is spirituality. There is an intrinsic force that creates and controls the universes we know, both within and without, and that force is omnipresent. Via our consciousness, we shall yield to the ultimate awareness. According to our religious values, we call the divine being by various names. Faith, with all its ups and downs, helps one to walk the road of life. It makes life happier for us.

A good awareness of SQ motivates people to align their job activities, time with family, and growth inside themselves. Thus, using this tutor, the student's potential can be viewed with sufficient comprehension and prepared to function properly to improve success during the course. The SQ analysis appears in Figures 10.14 and 10.15.

Name:____________________ **Division:**_____ **Branch:**_____ **Roll No:**______

Test: Q4 **Duration: 15 minutes.** **Date:** 09th July 2019

Circle the correct option

1. I experience joy in my studies.

   Strongly Disagree 1 2 3 4 5 6 7 Strongly Agree

2. I believe others experience joy as a result of my studies.

   Strongly Disagree 1 2 3 4 5 6 7 Strongly Agree

3. My spirit is energized by my study.

   Strongly Disagree 1 2 3 4 5 6 7 Strongly Agree

4. The study I do is connected to what I think is important in life.

   Strongly Disagree 1 2 3 4 5 6 7 Strongly Agree

5. I look forward to coming to study most days.

   Strongly Disagree 1 2 3 4 5 6 7 Strongly Agree

6. I see a connection between my study and the larger social good of my community.

   Strongly Disagree 1 2 3 4 5 6 7 Strongly Agree

FIGURE 10.14 Sample question of Spiritual quotient (SQ).

15 Questions
15 Minutes

7 point Likert-type scale
1- Strongly Disagree
7- Strongly Agree

Meaning of studies	(07)
Personal Responsibility	(02)
Positive Connection with other individuals	(01)
Conditions for Community	(01)
Inner Life	(02)
Institute and Individual	(02)

FIGURE 10.15 Sample Pattern of Spiritual quotient (SQ).

Above all are the 4QS model quotient uses for predicting the student performance in the coming semester so we can try to overcome the student issue for improving their result. We are also trying to use the stress quotient for predicting the performance with the 4QS model.

5) **Stress Analysis:** Stress, in our day-to-day terms, is an emotion that people face when they are highly loaded and experience difficulties while fulfilling daily demands. Short-term tensity can be beneficial; however, long-term tensity affects individual health to a considerable extent, such as cardiovascular disease, heart disease, high blood pressure, heart attack, and stroke. It can also lead to depression, anxiety, and personality disorders. Consequently, stress recognition becomes helpful to control health-related issues generated from tensity. Stress can be measured and evaluated dependent on perceptual, conduct, and physiological reactions. A few researchers have proposed different methodologies based on feature extraction and classification techniques. It is based on the fact that some of these techniques are intricate in their applicability and give less precise outcomes in human stress analysis.

Tensity is used to check how much stress students are facing while solving the question and doing the normal study so we can predict that properly. Following are some sample questions on Stress (see Figure 10.16).

Stress Management

Survey Questionnaire Name (optional)

:____________________________ Section: ___________ Gender: _______________ Age: _______

Q1. How often do you feel stressed?

( ) always ( ) everyday( ) sometimes ( ) once in a while( ) I don't know ( ) never 2.

Q2. In what aspect of life, you have the most problems?

( ) college ( ) family( ) friends ( ) love life( ) social life/community ( ) I don't know( ) others, please specify

____

FIGURE 10.16 Sample question of Stress Analysis

10.6 DISCUSSION AND RESULT

10.6.1 Discussion

The following section gives the detailed overview of the proposed 4QS model with education and curriculum development on different Quotient, 4QS Test Conduction (see Tables 10.3 and 10.4 and Figure 10.17):

- Written tests were conducted based on the 4QS model to test the quotients of students at the beginning.
- Questioners were shared with college students in the vicinity of Mumbai.
- Data collected includes deemed university, government-aided college, and minority college.

TABLE 10.3
Test Pattern with their details

Sr. No.	Quotient Tested	Number of Questions	Time Duration
Q1	Adversity Quotient	20	30 mins
Q2	Intelligence Quotient	30	60 mins
Q3	Emotional Quotient	15	15 mins
Q4	Spiritual Quotient	15	15 mins
S	Stress Test	10	10 mins

10.6.2 Test Conduction on Some Sample Dataset

TABLE 10.4
Different Branch college students Collected Data

College	Branch	Intake	Responses	Final	Total
C1	CMPN	60	59	58	
C1	IT	60	59	55	
C1	MECH	60	58	57	215
C1	MTRX	60	45	45	
C2	CHEM	60	30	28	
C2	CIVIL	60	47	45	129
C2	MECH	60	57	56	
C3	CMPN	60	52	50	
C3	IT	60	57	49	99
C4	CIVIL	120	52	31	
C4	CMPN	180	49	38	114
C4	MECH	120	46	45	

10.6.3 Results

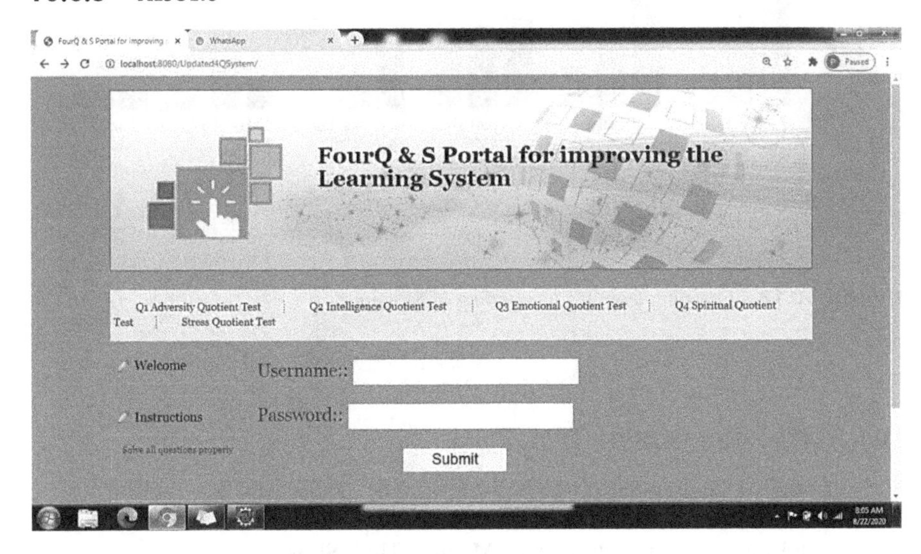

FIGURE 10.17 4QS Portal front page of Learning system

10.6.4 Identifying Students' Level of Understanding

Levels of a student (see Table 10.5 and Figure 10.18).

TABLE 10.5
Learning levels of a student

Roll No	Remembering IQ	Understanding IQ	Applying IQ,SQ	Analysing IQ, sa, EQ	Evaluating IQ, sa, EQ, AQ	Creating 4QS
##	H	H	M	M	M	M
##	H	H	H	H	M	H
##	H	H	H	M	M	M
##	M	M	M	M	M	L
##	L	L	M	L	L	L

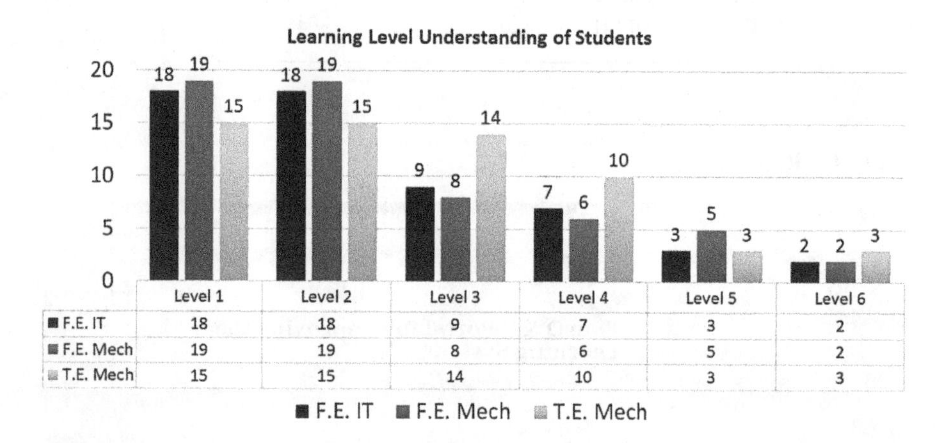

FIGURE 10.18 Learning level Understanding of Students.

10.6.5 Findings from the Test

- Stress was Medium and High in the F.E. students (Tables 10.6–10.10)
- Students with Medium and High AQ had less stress
- High Emotional students showed Medium-High Stress
- HSC Marks Intelligence Quotient did not match
- High Spiritual students showed less stress

TABLE 10.6
Prediction of Test 2 marks of FE students using KNN model for predicting their outcome

Cognitive Level	Students in each Level	Test 1 Percentage weightage	Prediction	Actual
Remember	18	10	Initial Prediction shows 5 students out of 57 will fail	3 Number of failures
Understand	18	10		
Apply	9	80		
Analyse	7	0		
Evaluate	3	0		
Create	2	0		

TABLE 10.7
Prediction of Test 2 marks of TE students using KNN model for predicting their outcome

Cognitive Level	Students in each Level	Test 1 Percentage weightage	Prediction	Actual
Remember	15	0	Initial prediction shows 15 students out of 60 will fail	10 number of failures
Understand	15	10		
Apply	14	50		
Analyse	10	30		
Evaluate	3	0		
Create	3	0		

TABLE 10.8
Prediction of Test 3 marks of SE students using KNN model for predicting their outcome

Cognitive Level	Students in each Level	Test 1 Percentage weightage	Prediction	Actual
Remember	18	10	Initial prediction shows 2 students out of 57 will fail	2 number of failures
Understand	18	10		
Apply	9	80		
Analyse	7	0		
Evaluate	3	0		
Create	2	0		

TABLE 10.9
Prediction of Test 3 marks of TE students using KNN model for predicting their outcome

Cognitive Level	Students in each Level	Test 1 Percentage weightage	Prediction	Actual
Remember	15	0	Initial prediction shows 09 students out of 60 will fail	11 number of failures
Understand	15	10		
Apply	14	40		
Analyse	10	50		
Evaluate	3	0		
Create	3	0		

TABLE 10
Accuracy prediction for FE and TE students by considering sample test

Test	Prediction		Actual		Difference	
	F.E. I.T.	T.E. MECH	F.E. I.T.	T.E. MECH	F.E. I.T.	T.E. MECH
T1	15	13	5	1	10	12
T2	5	15	3	10	2	5
T3	2	9	2	11	0	-2

Following are some points that are covered on some sample datasets (Figure 10.19):

- Four quotients with Stress to assess student's potential during course enrollment have been included in the lean model.
- Report based on 4QS and Cognitive Level indicating areas of improvement will help the student improve his/her performance.
- Teacher will be able to know the potential of the student before commencement of the semester.
- In curriculum planning, the Course Outcome can now be placed exactly with the cognitive level of students.
- Assessment pattern can now be mapped with the cognitive level for the proper level of teaching and learning.

Also, the knowledge and tools contained in this section would enable teachers and instructional administrators to see how lean concepts can be implemented to enhance the quality of the program and standardize current processes.

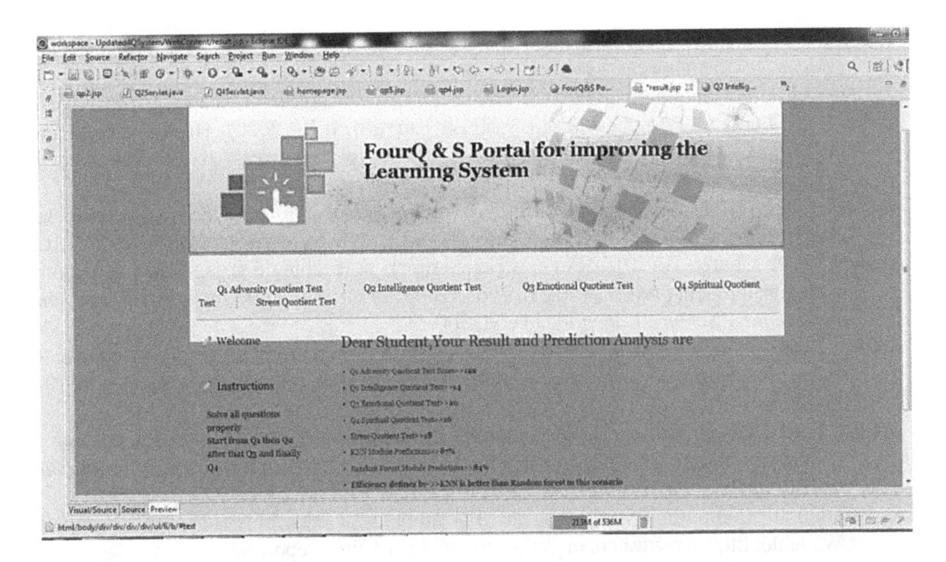

FIGURE 10.19 Overall Accuracy of 4QS Model using KNN and Random Forest Module.

10.7 CONCLUSION

Starting with a detailed analysis of the literature on lean principles and using lean philosophy in an academic environment, this chapter addressed the instance of an Education system in which lean principles were applied. Efficient application of lean principles contributed to the identification of the obsolete procedures and elements of the remaining curriculum. A new streamlined solution based on the pull-system was introduced. Lean will create a win-win situation in higher education institutions for curriculum growth for students, teachers, employers, and administrators by continuously concentrating on can-non-value—added initiatives. Furthermore, when applied correctly to core academic processes, the lean theory can contribute to the reduction or elimination of issues that lead to customer dissatisfaction. Applying lean methods would help create better services for both students and college/university as academic institutions are working to establish online classes. The knowledge and tools contained in this chapter would enable teachers and instructional administrators to see how lean concepts can be implemented to enhance the quality of the program and standardize current processes.

REFERENCES

Alves, A. C., Flumerfelt, S., & Kahlen, F. J. (Eds.). (2016). *Lean education: An overview of current issues*. Springer.

Alves, A. C., Flumerfelt, S., & Kahlen, F. J. (2017). *Lean education: An overview of current issues*. Springer International Publishing.

Alves, A. C., Kahlen, F. J., Flumerfelt, S., & Siriban-Manalang, A. B. (2013). *Lean Engineering Education: Bridging-the-gap between academy and industry*. Retrieved from https://repositorium.sdum.uminho.pt/bitstream/1822/30297/1/ShortVersionPaper_CISPEE.p df.

Alves, A. C., Leão, C. P., Uebe-Mansur, A. F., & Kury, M. I. R. (2020). The knowledge and importance of Lean Education based on academics' perspectives: An exploratory study. *Production Planning & Control*, 1–14.

Anderson, D. J. (2010). *Kanban: successful evolutionary change for your technology business*. BlueHolePress. [Online]. Available: https://books.google.es/books?id=RJ0VUkfUWZkC.

Anslow, C., & Maurer, F. (2015). An experience report at teaching a group based agile software development project course In *Proceedings of the 46th ACM Technical Symposium on Computer Science Education*. ACM, 500–505.

Antony, J. (2014). Readiness factors for the Lean Six Sigma journey in the higher education sector. *International Journal of Productivity and Performance Management, 63*(2), 257–264.

Antony, J., Krishan, N., Cullen, D., & Kumar, M. (2012). Lean Six Sigma for higher education institutions (HEIs): Challenges, barriers, success factors, tools/techniques. *International Journal of Productivity and Performance Management*, *6*(8), 940–948.

Association of Computing Machinery. 2013. Computer science curricula 2013. [Online]. Available: https://www.acm.org/education/CS2013-final-report.pdf.

Badawy, M. (2006). Preparing engineers for management. *Engineers Australia*, *7*, 64–66.

Baillie, C., & Fitzgerald, G. (2000). Motivation and attrition in engineering students. *European Journal of Engineering Education*, 25(2), 145–155.

Balaji, K. V., & Somashekar, P. (2009). A comparative study of soft skills among engineers. *IUP Journal of Soft Skills*, 3(3–4), 50–57.

Balzer, W. (2010). *Lean higher education: increasing the value and performance of university processes*. CRC Press.

Balzer, W., Brodke, M., & Kizhakethalackal, E. T. (2015). Lean higher education: Successes, challenges, and realizing potential. *International Journal of Quality & Reliability Management*, 32(9), 924–933.

Barling, J. (2014). *The science of leadership*. Oxford University Press.

Beck, K. (2000). Extreme programming explained: Embrace Change, ser. An Alan R. Apt Book Series. Addison-Wesley. [Online]. Available: https://books.google.co.uk/books?id=G8EL4H4vf7UC.

Berdanier, C., Branch, S., London, J., Ahn, B., & Farmer, M. (2014). *Survey Analysis of Engineering Graduate Students' Perceptions of the Skills Necessary for Career Success in Industry and Academia, 121st ASEE Annual Conference and Exposition*, June 15–18, Indianapolis.

Bishop, J. H. (1995). Expertise and Excellence, (CAHRS Working Paper #95–13), Ithaca, NY: Cornell University, School of Industrial and Labor Relations, Center for Advanced Human Resource Studies, Retrieved from http://digitalcommons.ilr.cornell.edu/cgi/viewcontent.cgi?article=1202&context=cahrswp.

Bol, L. A. (2010). *Knowledge, skills, and abilities necessary for success in the manufacturing industry*. University of Wisconsin-Stout Menomonie.

Byrne, A. (2013). *Lean turnaround: How business leaders use lean principles to create value and transform their company*. McGraw-Hill.

Carvalho, C. V., Lopes, M. P., António, G., Ávila, P., Bastos, J., & Fonseca, L. (2013). *Lean learning academy: an innovative framework for lean manufacturing training, 1st International Conference of the Portuguese Society for Engineering Education (CISPEE)*, October 31—November 1, Porto.

CBI. (2016). The right combination/Pearson education and skills survey 2016. Retrieved from http://www.cbi.org.uk/cbi-prod/assets/File/pdf/cbi-education-and-skills-survey2016.pdf.

Douglas, J. A., Antony, J., Ciasullo, M. V., & Douglas, A. (2020). Recognising waste in higher education institutions using lean thinking. *Lean six sigma for higher education: research and practice*, 21.

K-Nearest Neighbor(KNN) Algorithm for Machine Learning—Javatpoint. www.javatpoint.com. (2021). Retrieved, from https://www.javatpoint.com/k-nearest-neighbor-algorithm-for-machine-learning.

Kropp, M., & Meier, A. (2014). *New sustainable teaching approaches in software engineering education*. In *2014 IEEE Global Engineering Education Conference (EDUCON)*. IEEE. 1019–1022.

Machine Learning Random Forest Algorithm—Javatpoint. www.javatpoint.com. (2021). Retrieved, from https://www.javatpoint.com/machine-learning-random-forest-algorithm.

Nanade, S., & Lal, S. (2019a, May). Applying lean for effective implementation and governance of education in future cities. *International Journal of Innovative Technology and Exploring Engineering (IJITEE)*. *8*(7S) 2278–3075.

Nanade, S., & Lal, S. (2019b, June). Developing engineering curriculum: The lean way. *International Journal of Innovative Technology and Exploring Engineering (IJITEE)*, *8*(8S3) 2278–3075.

Nanade, S., & Lal, S., Goswami, S., & Sharma, A. (n.d.). 4QS model to assess the student learning capabilities. *International Journal of Advanced Science and Technology (IJAST)* 2227–2244. Retrieved from http://sersc.org/journals/index.php/IJAST/article/view/24405-2020.

Papatheocharous, E. & Andreou, A. S. (n.d.). Empirical evidence and state of practice of software agile teams. *Journal of Software: Evolution and Process*, *26*(9).

Index